T0211459

MICROBIOLOGY FOR
WATER/WASTEWATER OPERATORS

HOW TO ORDER THIS BOOK

BY PHONE: 800-233-9936 or 717-291-5609, 8AM–5PM Eastern Time

BY FAX: 717-295-4538

BY MAIL: Order Department
Technomic Publishing Company, Inc.
851 New Holland Avenue, Box 3535
Lancaster, PA 17604, U.S.A.

BY CREDIT CARD: American Express, VISA, MasterCard

BY WWW SITE. http://www.techpub.com

PERMISSION TO PHOTOCOPY—POLICY STATEMENT

Authorization to photocopy items for internal or personal use, or the internal or personal use of specific clients. is granted by Technomic Publishing Co.. Inc. provided that the base fee of US $3.00 per copy. plus US $.25 per page is paid directly to Copyright Clearance Center, 222 Rosewood Drive, Danvers. MA 01923, USA. For those organizations that have been granted a photocopy license by CCC. a separate system of payment has been arranged The fee code for users of the Transactional Reporting Service is 1-56676/00 $5.00 + $.25.

Microbiology for Water/Wastewater Operators

REVISED REPRINT

FRANK R. SPELLMAN, Ph.D.

CRC Press
Taylor & Francis Group
Boca Raton London New York

CRC Press is an imprint of the
Taylor & Francis Group, an **informa** business

CRC Press
Taylor & Francis Group
6000 Broken Sound Parkway NW, Suite 300
Boca Raton, FL 33487-2742

First issued in paperback 2019

© 2000 by Taylor & Francis Group, LLC
CRC Press is an imprint of Taylor & Francis Group, an Informa business

No claim to original U.S. Government works

ISBN-13: 978-1-56676-908-2 (hbk)
ISBN-13: 978-0-367-39915-3 (pbk)

This book contains information obtained from authentic and highly regarded sources. Reasonable efforts have been made to publish reliable data and information, but the author and publisher cannot assume responsibility for the validity of all materials or the consequences of their use. The authors and publishers have attempted to trace the copyright holders of all material reproduced in this publication and apologize to copyright holders if permission to publish in this form has not been obtained. If any copyright material has not been acknowledged please write and let us know so we may rectify in any future reprint.

Except as permitted under U.S. Copyright Law, no part of this book may be reprinted, reproduced, transmitted, or utilized in any form by any electronic, mechanical, or other means, now known or hereafter invented, including photocopying, microfilming, and recording, or in any information storage or retrieval system, without written permission from the publishers.

For permission to photocopy or use material electronically from this work, please access www.copyright.com (http://www.copyright.com/) or contact the Copyright Clearance Center, Inc.(CCC), 222 Rosewood Drive, Danvers, MA 01923, 978-750-8400. CCC is a not-for-profit organization that provides licenses and registration for a variety of users. For organizations that have been granted a photocopy license by the CCC, a separate system of payment has been arranged.

Trademark Notice: Product or corporate names may be trademarks or registered trademarks, and are used only for identification and explanation without intent to infringe.

**Visit the Taylor & Francis Web site at
http://www.taylorandfrancis.com**

**and the CRC Press Web site at
http://www.crcpress.com**

Main entry under title:
    Microbiology for Water/Wastewater Operators, Revised Reprint

A Technomic Publishing Company book
Bibliography: p. 199
Includes index p. 205

Library of Congress Catalog Card No. 99-68514

To all the water/wastewater specialists
throughout this water-shrouded planet

THIS book is designed to provide a comprehensive discussion of microbiological concepts as they relate to water and wastewater. This book assumes no prior knowledge of microbiology. From the first page on, it builds an integrated picture of the structure, growth, morphology, and metabolism of microorganisms; these fundamental concepts are central to all areas of microbiology—including bacteriology, water/wastewater treatment, and environmental science. Sufficient background information and reference to microbiological and bacteriological principles and practices are presented throughout the text to provide readers with an understanding of the various concepts and organisms under discussion. Subsequent chapters and the Appendix deal with practical applications of microbiology and also cover identification of microorganisms, aseptic technique, fecal-coliform isolation, identification, and verification. Moreover, operational troubleshooting hints are included for personnel who are concerned with the operational conditions existing in activated sludge systems.

This book can be used as a basic study tool by water/wastewater personnel who are preparing for their licensing examinations or as a supplemental text in undergraduate or graduate courses in aquatic ecology, water/wastewater pollution control, and environmental science courses dealing with water biology. It can also be consulted as a biological/environmental science reference text by municipal, school, and water/wastewater resource libraries.

This update augments previous information and emphasizes

the new world order of water control based on microbiological principles and practices. Emphasis has been placed on the new pathogen protozoa and their deleterious effects on the "sensitive population" (the infant, the chronically ill, elderly, and the AIDS population). It treats the important new parameters of *Cryptosporidium* and the enhanced surface water treatment requirements.

After having been displaced in importance in the 1970s and 1980s, microbiology (parameters and controls), returned to the forestage, its efficacy in water treatment reestablished. Disinfection was also back in favor, along with more sophisticated water treatment practices. Except for lead, arsenic, and radon, the fear of carcinogens was demoted from the top level of regulatory attention, and in its place *Cryptosporidium* and *Giardia* have become the targets of concern.

The preparation of this revision was helped by the comments of those who count the most: the water/wastewater practitioners in the field who perform their professional activities on a "daily" basis.

ACKNOWLEDGEMENTS

I would like to acknowledge the numerous contributions made to this book in the area of microbiological laboratory practices by the Water Quality Department of Hampton Roads Sanitation District (HRSD), Virginia Beach, Virginia. Without a doubt, the department has the finest group of wastewater laboratory specialists anywhere. For specific professional assistance I am also grateful to Robert Maunz, Laboratory Specialist A, and to Nancy Davis, Laboratory Technician, HRSD. Without their assistance, many of the practical applications addressed in this book would not have been correct. And, of course, without the help of my two associates, Jennifer Rodrigues and Rosalyn Hopkins, this book would not have been possible.

Finally, I am inspired to write something about environmental science whenever I look upon my granddaughter's (Rachel Morgan Spellman) shining, smiling face. I guess it is the innocence and helplessness that I really see. I look at her and wonder if she understands the place in time that we now occupy on earth in relation to our environment and if she and her friends will have a place to live where the water is clean—free of disease and pollution. Will she have the chance to grow to old age and gain the wisdom to understand that she was born of water and needs to be sustained by it?

I hope so!

FRANK R. SPELLMAN

PART 1

Fundamental Concepts

CHAPTER 1

Introduction

Water is sometimes sharp and sometimes strong, sometimes acid and sometimes bitter, sometimes sweet and sometimes thick or thin, sometimes it is seen bringing hurt or pestilence, sometimes health-giving, sometimes poisonous. It suffers change into as many natures as are the different places through which it passes. And as the mirror changes with the color of its object, so it alters with the nature of the place, becoming: noisome, laxative, astringent, sulfurous, salt, incarnadined, mournful, raging, angry, red, yellow, green, black, blue, greasy, fat or slim. Sometimes it starts a conflagration, sometimes it extinguishes one; is warm and is cold, carries away or sets down, hollows out or builds up, tears down or establishes, fills or empties, raises itself or burrows down, spreads or is still; is the cause at times of life or death, or increase or privation, nourishes at times and at others does the contrary; at times has a tang, at times it is without savor, sometimes submerging the valleys with great flood. In time and with water, everything changes.

Leonardo da Vinci

THROUGH experience, water and wastewater specialists come to know many of the characteristics of water described by da Vinci. It is, however, the possibility that water can "bring hurt and pestilence" to other organisms that most interests water and wastewater specialists.

Water treatment specialists are concerned with water supply and water purification through a treatment process. In treating water, the primary concern is producing potable water that is safe to drink (free of pathogens) with no accompanying offen-

3

sive characteristics such as foul taste and odor. The water specialist must possess a wide range of knowledge in order to correctly examine water for pathogenic microorganisms and to determine the type of treatment necessary to ensure that the water quality of the end product, potable water, meets regulatory standards.

Wastewater treatment specialists are also concerned with water quality. However, they are not as concerned as water specialists are with total removal or reduction of most microorganisms. The wastewater treatment process actually benefits from microorganisms that act to degrade organic compounds and, thus, stabilize the organic matter in the waste stream. Thus, wastewater specialists must be trained to operate the treatment process in a manner that controls the growth of microorganisms and puts them to work in the stabilization process. Moreover, to more fully understand wastewater treatment, it is necessary to determine which microorganisms are present and how they function to break down the components in the wastewater stream. Then, of course, the wastewater specialist must ensure that before dumping treated effluent into a receiving body, the microorganisms that worked hard to degrade organic waste products, especially the pathogenic microorganisms, are not sent from the plant as viable organisms.

The average citizen living in the United States or Europe has heard of waterborne disease-causing microorganisms, but in this modern age he or she probably does not give them a second thought. Modern sanitation practices have made contraction of most of the waterborne diseases, shown in Table 1.1, rare in the United States and Europe. This is not the case, however, in other areas of the world. It would be foolhardy (and deadly) for us to forget that disease-causing organisms are still in our environment.

In the water environment, Koren (1991) points out that water is not a medium for the *growth* of microorganisms, but is instead a means of transmission (a conduit for; hence, the name *waterborne*) of the pathogen to the place where an individual is able to consume it and there start the outbreak of disease. This is contrary to the view taken by the average person. That is, when the topic of waterborne disease is brought to his/her at-

Table 1.1. Waterborne disease-causing organisms.

Microorganism	Disease
Bacterial	
Escherichia coli	
Salmonella typhi	Typhoid fever
Salmonella spl.	Salmonellosis
Shigella sp.	Shigellosis
Yersinia entercolitica	Yersiniosis
Vibrio cholerae	Cholera
Campylobacter jejuni	Campylobacter enteritis
Intestinal parasites	
Entamoeba histolytica	Amebic dysentery
Giardia lamblia	Giardiasis
Cryptosporidium	Cryptosporidiosis
Viral	
Norwalk agent	—
Rotavirus	—
Enterovirus	Polio
	Aseptic meningitis
	Herpangina
Hepatitis A	Infectious hepatitis
Adenoviruses	Respiratory disease
	Conjunctivitis

tention, he/she might mistakenly assume that waterborne diseases are at home in water. Nothing could be further from the truth. A water-filled *ambience* is not the environment in which the pathogenic organism would choose to live, that is, if it had such a choice. The point is that microorganisms do not normally grow, reproduce, languish, and thrive in watery surroundings. Pathogenic microorganisms temporarily residing in water are simply biding their time, going with the flow, waiting for their opportunity to meet up with their unsuspecting host or hosts. To a degree, when the pathogenic microorganism finds its host or hosts, it is finally home or may have found its final resting place.

Water treatment operations generally focus on operating, monitoring, and determining settings for chemical feed machines and high pressure pumps and boilers. Likewise, wastewater treatment operations are also concerned with the general aspects of process flow treatment and management. The important point is that neither of these water treatment specialties can accomplish their missions without having a well-rounded, fundamental knowledge of the science of microbiology. This is important. For example, the water specialist who is well grounded in microbiological concepts is equipped to operate his or her plant in a manner that will provide safe, sanitary, palatable quality drinking water for domestic consumption. Likewise, the wastewater specialist who has knowledge of microbiological concepts is equipped to operate wastewater treatment plant processes in a manner that will produce effluent of better quality (hopefully) than the water contained in the receiving body.

As related above, it would be impossible for the water or wastewater specialist to fully comprehend the principles of effective water/wastewater treatment without having knowledge of the fundamental factors concerning microorganisms and their relationships to one another, their effect on the treatment process, and their impact on the environment, human beings, and other organisms. Thus, the intention of this text is to provide a fundamental knowledge of microbiology for water and wastewater specialists. In order to provide this fundamental knowledge, we will pursue a structured approach that is basic but far reaching.

MICROBIOLOGY

Microbiology is the study of organisms that are of microscopic dimensions and thus cannot be seen except with the aid of a microscope. Microbiologists are scientists who are concerned with studying the form, structure, reproduction, physiology, metabolism, and identification of microorganisms. The microorganisms they study generally include bacteria, fungi, protozoa, algae, and viruses. These tiny organisms make up a large and diverse group of free-living forms that exist either as single cells, cell bunches or clusters. Any and all of these organisms may be found in water and/or wastewater.

Microscopic organisms can be found in abundance almost anywhere on earth. The vast majority of microorganisms are not harmful. Many microorganisms, or microbes, occur as single cells (unicellular); others are multicellular; and still others, viruses, do not have a true cellular appearance.

Because microorganisms exist as single cells or cell bunches, they are unique and distinct from the cells of animals and plants, which are not able to live alone in nature but can exist only as part of multicellular organisms (Brock & Madigan, 1991). A single microbial cell, for the most part, exhibits the characteristic features common to other biological systems, such as metabolism, reproduction, and growth.

CLASSIFICATION

For centuries, scientists classified the forms of life visible to the naked eye as either animal or plant. Much of the current knowledge about living things was organized by the Swedish naturalist Carolus Linnaeus in 1735.

The importance of classifying organisms cannot be overstated, for without a classification scheme, it would be difficult to establish a criteria for identifying organisms and to arrange similar organisms into groups. Probably the most important reason for classifying organisms is to make things less confusing (Wistreich & Lechtman, 1980).

Linnaeus was quite innovative in the classification of organisms. One of his innovations is still with us today: *the binomial system of nomenclature.* Under the binomial system all organisms are generally described by a two-word scientific name, the *genus* and *species.* Genus and species are groups that are part of a hierarchy of groups of increasing size, based on their nomenclature (taxonomy). This hierarchy follows.

Kingdom

Phylum

Class

Order

Family

Genus

Species

Utilizing this hierarchy and Linnaeus's binomial system of nomenclature, the scientific name of any organism (as stated previously) includes both the genus and the species name. The genus name is always capitalized, while the species name begins with a lowercase letter. On occasion, when there is little chance for confusion, the genus name is abbreviated with a single capital letter. The names are always in Latin, so they are usually printed in italics or underlined. Some organisms also have English common names. Some microbe names of interest in water/wastewater treatment follow.

- *Salmonella typhi*—the typhoid bacillus
- *Escherichia coli*—a coliform bacteria
- *Giardia lamblia*—a protozoan

Escherichia coli is commonly known as simply *E. coli*, while *Giardia lamblia* is usually referred to by only its genus name, *Giardia*.

A simplified system of microorganism classification is used in water and wastewater. Classification is broken down into the kingdoms of animal, plant, and protista. As a general rule the animal and plant kingdoms contain all the multicell organisms, and the protists contain all single-cell organisms. Along with microorganism classification based on the animal, plant, and protista kingdoms, microorganisms can be further classified as

Table 1.2. Simplified classification of microorganisms.

Kingdom	Members	Cell Classification
Animal	Rotifers Crustaceans Worms and larvae	
		Eucaryotic
Plant	Ferns Mosses	
Protista	Protozoa Algae Fungi	
	Bacteria Lower algae forms	Procaryotic

being *eucaryotic* or *procaryotic* (see Table 1.2). A eucaryotic organism is characterized by a cellular organization that includes a well-defined nuclear membrane. A procaryotic organism is characterized by a nucleus that *lacks* a limiting membrane.

THE CELL

Since the nineteenth century, scientists have known that all living things, whether animal or plant, are made up of cells. Moreover, the fundamental unit of all living matter, no matter how complex, is the cell. A typical cell is an entity, isolated from other cells by a membrane or cell wall. The cell membrane contains protoplasm, the living material found within it, and the nucleus. In a typical mature plant cell, the cell wall is rigid and is composed of nonliving material, while in the typical animal cell, the wall is an elastic living membrane. Cells exist in a very great variety of sizes and shapes, as well as functions. Their average size ranges from bacteria too small to be seen with the light microscope to the largest known single cell, the ostrich egg. Microbial cells also have an extensive size range, some being larger than human cells (Kordon, 1993). Much more will be said about the cell later in this text.

SUMMARY OF KEY TERMS

- *Microbiology:* may be defined in terms of the size of the organisms studied and techniques employed
- *Procaryotic cells:* differ from *eucaryotic* cells in that the former lack a membrane-delimited nucleus and in other ways.

CHAPTER 2

The Microscope

SINCE about 1680, when van Leeuwenhoek first observed microbes in his simple microscope, the microscope has assumed a central role in the science of microbiology. The microscopes used today have evolved significantly from Leeuwenhoek's first microscope. Depending upon the magnification principle involved, microscopes are known as either light microscopes or electron microscopes. The light, or optical, microscope is the type that is most commonly used in water and wastewater treatment and is discussed here.

In water and wastewater treatment two types of light microscopes are generally used: a single-lens system for high magnification and a two-lens or binocular (stereoscopic) microscope for lower magnification (see Figure 2.1). Certain principles underlie the operation of all microscopes.

1. *Magnification* is accomplished by the use of one or more lenses. In a *compound microscope,* the lens nearest the object (objective lens) magnifies, producing a real image. The eyepiece (ocular lens) magnifies the real image, producing a virtual image seen by the eye. Microscopes generally consist of three or more objective lenses. The low-power objective lens is about 10 power, or 10× (10× is a ten times increase in size). The high-power lens generally is 40×. Most microscopes are equipped with a third lens of about 100×. The 100× lens is an oil-immersion lens; it magnifies through oil between the lens and the specimen. Because a 100- or 400-power lens is typically used in water/wastewater labs, the oil-immersion lens is an especially important tool. The best results are obtained when the objective is immersed in oil having approximately the same in-

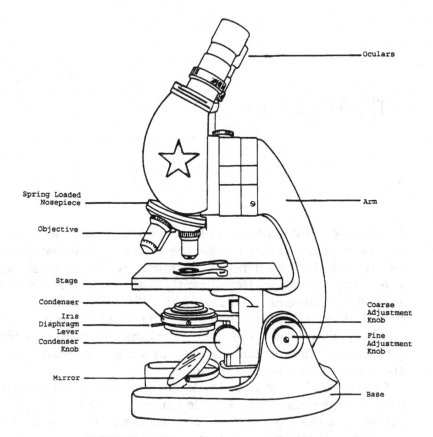

FIGURE 2.1 *Binocular (stereoscopic) microscope.*

dex of refraction as glass, about 1.6. Oils have the advantage of not evaporating when exposed to air for long periods of time. Moreover, since its index of refraction is the same as that of glass, oil does not bend the light rays entering the front lens of the oil-immersion objective.

The key point to remember about microscope magnification is that the total magnification of the microscope is the product of the magnification of the objective and the magnification of the eyepiece. That is

magnification objective × magnification eyepiece

= total magnification

2. McKinney (1962) points out that the sensitivity of the microscope is affected by more than just its ability to magnify an object. The microscope's *resolution*, or *resolving power*, is also important. Resolution or resolving power is the ability of a microscope to distinguish between two closely adjacent points. The resolution of a microscope lens system is a function of its numerical aperture and the wavelength of light (shorter wavelengths provide finer detail). *Numerical aperture* (NA) (usually engraved on the barrel of each objective by the manufacturer) refers to the ability of a lens to gather light and is expressed by the relation

$$NA = i \sin \theta \tag{2.1}$$

where

θ = one-half the angle of the entering cone of light
i = the index of refraction of the medium along the light path

The index of refraction for air is 1.0. In order to increase the light-gathering ability of the lens, something with an index of refraction greater than 1.0 must be placed between the object and the lens.

Resolution or resolving power can be calculated as follows:

$$d = \frac{\lambda}{2(NA)} \tag{2.2}$$

where

d = diameter of the smallest observable object (μm)
λ = wavelength of light (μm)
NA = numerical aperture (unitless)
2 = the use of a condenser below the stage to collect light to illuminate the specimen (Kelley & Post, 1989)

3. *Working distance* is the space between the objective and the specimen when the specimen is in focus. The magnification and the working distance are directly related. That is, the

higher the magnification, the shorter the distance. Generally speaking, the shorter the working distance, the greater the amount of light required to see the specimen.

STAINS

Microbial cells are nearly transparent when observed by light microscopy and hence are difficult to see. The most common method for observing cells is by the use of stained preparations. Dyes are used to stain cells. Staining cells increases their contrast so that they can be more easily observed in the light microscope. Some of the most commonly used dyes for staining are basic fuchsin, crystal violet, and methylene blue. These dyes work particularly well on bacteria because they have chromophores (color-bearing ions) that are cationic (positively charged). The fact that bacteria are negatively charged produces a pronounced attraction between these cationic chromophores and the organism. Figure 2.2 shows the steps to follow for staining cells for microscopic observation.

Spread culture in thin film over slide

Dry in air

Pass slide through flame to fix

Flood slide with stain; rinse and dry

Place drop of oil on slide;
examine with 100X objective

FIGURE 2.2 Staining procedure.

GRAM STAIN

The simple staining technique previously described depends upon the fact that bacterial cells differ chemically from their surroundings and thus can be stained to contrast with their environment. Microbes also differ from one another chemically and physically and, therefore, may react differently to a given staining procedure. This is the basic principle of *differential staining*—so named because this type of procedure does not stain all kinds of cells equally.

The Gram staining procedure was developed in the 1880s by Hans Christian Gram, a Danish bacteriologist. While attempting to stain pathologic specimens, Gram discovered that microbes could be distinguished from surrounding tissue. Gram observed that some bacterial cells exhibit an unusual resistance to decolorization. He used this observation as the basis for a differential staining technique.

The Gram differentiation is based upon the application of a series of four chemical reagents: *primary stain*, *mordant*, *decolorizer*, and *counterstain*. The purpose of the primary stain, crystal violet, is to impart a blue or purple color to all organisms regardless of their designated Gram reaction. This is followed by the application of Gram's iodine, which acts as a mordant (fixer) by enhancing the union between the crystal violet stain and its substrate by a forming complex. The decolorizing solution of 95% ethanol extracts the complex from certain cells more readily than others. In the final step a counterstain, safranin, is applied in order to see those organisms previously decolorized by removal of the complex. Those organisms retaining the complex are gram-positive (blue or purple), whereas those losing the complex are gram-negative (red or pink).

Undoubtedly, Gram staining is the most important staining procedure you will use in the identification of unknown bacteria. This technique is rather simple, but performing it with a high degree of reliability requires practice and experience. In order to increase the reliability of your Gram stains remember to make your stains thin, not too thick. Secondly, be careful when you decolorize with alcohol; it is important to allow 10–20 seconds for the solvent to flow colorlessly. Finally, avoid using old cultures of gram-positive bacteria.

Spore Stain

Bacterial spores are known for their resistance to high temperature, drying, chemical disinfection, and staining. Species of the gram-positive genera *Bacillus* and *Clostridium* produce heat-resistant structures referred to as endospores (which are formed within the cell). A thick, tough spore coat enables them to resist heat. These structures cannot be stained by ordinary methods. The stains do not penetrate the spore's wall. Stained smears of spore-containing cells appear to have oval holes, or colorless spheres, within them.

Several procedures are available that employ heat to provide stain penetration. However, two methods, the Schaeffer-Fulton and Dorner procedures, are the most commonly used procedures and will be described here.

In the Schaeffer-Fulton procedure, the primary stain, malachite green, is applied to a heat-fixed smear and heated to steaming over a beaker of boiling water for approximately 5 minutes. After the slide has cooled sufficiently, the slide is rinsed with water for about 30 seconds. Then the counterstain safranin is applied for about 20 seconds. Safranin displaces any residual dye in the nonsporulating cells but not in the spores. The safranin is then removed by rinsing the slide with water. Finally, the slide is blot dried and examined under oil immersion (Singleton & Sainsbury, 1994).

The Dorner method for staining endospores produces a red spore within a colorless sporangium. Nigrosine is used to provide a dark background for contrast. The Dorner staining method begins with making a heavy suspension of bacteria by dispersing several loopfuls of bacteria in five drops of sterile water and adding five drops of carbolfuchsin to the bacterial suspension. Heat is applied to the carbolfuchsin suspension of bacteria in a beaker of boiling water for 10 minutes. Although both the sporangium and endospore are stained during this procedure, the sporangium is decolorized by the diffusion of safranin molecules into the nigrosine. The next step requires the mixing of several loopfuls of bacteria in a drop of nigrosine on the slide. Spread the nigrosine-bacteria mixture on the slide by moving a spreader slide toward the drop of suspension until it contacts the drop causing the liquid to be spread along the spreading edge. Finally, allow the smear to air dry and examine the slide under oil immersion.

SUMMARY OF KEY TERMS

- *Compound Microscope:* the primary image is formed by an objective lens and enlarged by the eyepiece or ocular lens to yield a virtual image.
- *Resolution:* increases as the wavelength of radiation used to illuminate the specimen decreases; the maximum resolution of a light microscope is about 0.2 μm.
- *Staining:* specimens usually must be stained before viewing them in the bright-field microscope.
- *Staining dyes:* either positively charged basic dyes or negative acid dyes that bind to ionized parts of cells.
- *Differential staining:* the Gram stain and acid-fast stain distinguish between microbial groups by staining them differently.

CHAPTER 3

Bacteria

OF all the microorganisms studied in this text, bacteria are the most widely distributed, the smallest in size, the simplest in morphology (structure), the most difficult to classify, and the hardest to identify. Because of considerable diversity, it is even difficult to provide a descriptive definition of a bacterial organism. About the only generalization that can be made is that bacteria are single-celled plants, are procaryotic, are seldom photosynthetic, and reproduce by binary fission.

Bacteria are found everywhere in our environment. They are present in the soil, in water, and in the air. Bacteria are also present in and on the bodies of all living creatures, including man. Most bacteria do not cause disease; that is, they are not pathogenic. Many bacteria carry on useful and necessary functions related to the life of larger organisms.

When we think about bacteria in general terms, we usually think of the damage they cause. For example, Black-Covilli (1992) points out that "the form of water pollution that poses the most direct menace to human health is bacteriological contamination" (p. 23). This is partly the reason that bacteria are of great significance to water and wastewater specialists. For water treatment personnel tasked with providing the public with safe, potable water, disease-causing bacteria pose a constant challenge (see Table 3.1).

Wastewater specialists and water treatment specialists share the challenge of controlling pathogenic bacteria. Domestic wastewater normally contains huge quantities of microorganisms, such as bacteria, viruses, protozoa, and worms.

Table 3.1. Disease-causing bacterial organisms found in polluted water.

Microorganism	Disease
Salmonella typhi	Typhoid fever
Salmonella sp.	Salmonellosis
Shigella sp.	Shigellosis
Campylobacter jejuni	Campylobacter enteritis
Yersinia entercolitica	Yersiniosis
Escherichia coli	

Even though wastewater can contain bacteria counts in the millions per ml, in wastewater treatment, under controlled conditions, bacteria can help to destroy pollutants (McGhee, 1991). In such a process, bacteria stabilize organic matter (e.g., activated sludge processes) and thereby assist the treatment process in producing effluent that does not impose an excessive oxygen demand on the receiving body (Kemmer, 1979).

HOW WELL DO WE KNOW BACTERIA?

The conquest of disease has placed bacteria high on the list of microorganisms of great interest to the scientific community. It must be pointed out, however, that there is more to this interest and large research effort than just an incessant search for understanding and the eventual conquest of disease-causing bacteria. As stated earlier, not all bacteria are harmful to man. Some, for example, produce substances (antibiotics) that help in the fight against disease, while others are used to control insects that attack crops. Bacteria also affect the natural cycles of matter. For example, bacteria work to increase soil fertility, which, in turn, increases the potential for more food production. With the burgeoning world population, increasing future food productivity is no small matter.

We still have a great deal to learn about bacteria. Thomas (1983) points out that we are "still principally engaged in making observations and collecting facts, trying wherever possible to relate one set of facts to another but still lacking much of a basis for grand unifying theories" (p. 71). Like most learning

processes, gaining knowledge about bacteria is a slow and delib-
erate process. There can be little doubt, however, that the more
we know about bacteria, the more we can minimize their harmful potential and exploit their useful activities.

SHAPES, FORMS, SIZES, AND ARRANGEMENTS OF BACTERIAL CELLS

Bacteria come in three shapes: elongated rods called *bacilli*,
rounded or spherical cells called *cocci*, and spirals (helical and
curved) called *spirilla* (for the less rigid form) and *spirochaete*
(for those that are flexible). Elongated rod-shaped bacteria may
vary considerably in length; have square, round, or pointed
ends, and be motile (able to move) or nonmotile. The spherical-
shaped bacteria may occur singly, in pairs, in tetrads, in chains,
and in irregular masses. The helical and curved spiral-shaped
bacteria exist as slender spirochaetes, spirillum, and bent rods
(see Figure 3.1).

Bacterial cells are usually measured in microns, μ, or micrometers, μm; 1 μm = 0.001 or 1/1,000 of a millimeter, mm. A
typical rod-shaped coliform bacterial cell is about 2 μm long and
about 0.7 microns wide. The size of a cell changes with time
during growth and death.

Bacterial cells, viewed under the microscope, may be seen as
separate (individual) cells or as cells in groupings. According to
the species, cells may appear in pairs (diplo), chains, groups of
four (tetrads), cubes (sarcinae), and clumps. Long chains of
cocci result when cells adhere after repeated divisions in one
plane; this pattern is seen in the genera *Enterococcus* and
Lactococcus. In the genus *Sarcina*, cocci divide in three planes
producing cubical packets of eight cells (tetrads). The exact
shape of rod-shaped cells varies, especially at the end of the rod.
The rod's end may be flat, cigar shaped, rounded, or bifurcated.
While many rods do occur singly, they may remain together after division to form pairs or chains (see Figure 3.1). All of these
characteristic arrangements are frequently useful in bacterial
identification.

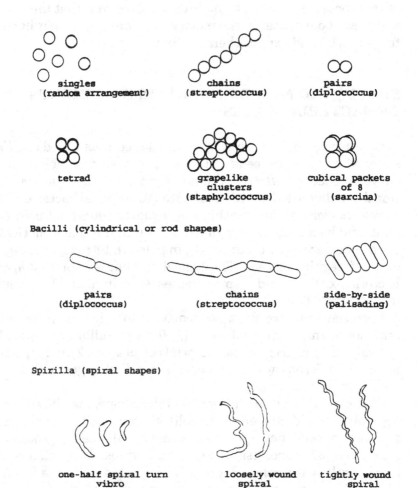

FIGURE 3.1 *Bacterial shapes and arrangements.*

BACTERIAL CELL SURFACE STRUCTURES AND CELL INCLUSIONS

Cell structure can best be studied in the rod form as shown in Figure 3.2. One point to keep in mind when studying Figure 3.2 is that cells of different species may differ greatly both in their

structure and chemical composition; for this reason there is no
typical bacterium. Again, Figure 3.2 shows a generalized bacterium; it is important to note that not all bacteria have all the
features shown in the figure and that some bacteria have structures not shown in the figure.

Capsule

Bacterial capsules (see Figure 3.2) are organized accumulations
of gelatinous material on cell walls in contrast to *slime layers* (a
water secretion that adheres loosely to the cell wall and commonly diffuses into the cell), which are unorganized accumulations of similar material. The capsule is usually thick enough to be
seen under the ordinary light microscope (macrocapsule), while
thinner capsules (microcapsules) can be detected only by electron
microscopy (Singleton & Sainsbury, 1994).

The production of capsules is determined largely by genetics,

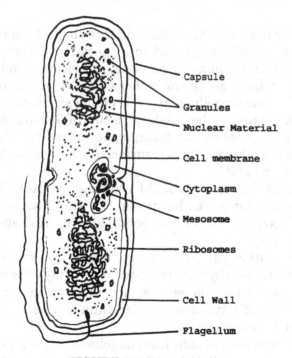

FIGURE 3.2 *Bacterial cell.*

as well as environmental conditions, and depends on the presence or absence of capsule-degrading enzymes and other growth factors. Varying in composition, capsules are mainly composed of water; the organic contents are made up of complex polysaccharides, nitrogen-containing substances, and polypeptides.

Capsules confer several advantages when bacteria grow in their normal habitat. For example, they help (1) to prevent desiccation; (2) bacteria resist phagocytosis by host phagocytic cells; (3) prevent infection by bacteriophages; and (4) aid bacterial attachment to tissue surfaces in plant and animal hosts or to surfaces of solid objects in aquatic environments. Capsule formation often correlates with pathogenicity. Capsule-secreted polysaccharides have been used for industrial purposes. In the food industry, for example, the polysaccharides have been used as gelling agents (Singleton, 1992).

Flagella

Many bacteria are motile, and this ability to move independently is usually attributed to a special structure, the *flagella* (singular: flagellum). Depending on species, a cell may have a single flagellum (*monotrichous* bacteria; *trichous* means "hair"); one flagellum at each end (*amphitrichous* bacteria; *amphi* means "on both sides"); a tuft of flagella at one or both ends (*lophotrichous* bacteria; *lopho* means "tuft"); or flagella that arise all over the cell surface (*peritrichous* bacteria; *peri* means "around").

A flagellum is a threadlike appendage extending outward from the plasma membrane and cell wall. Flagella are slender, rigid locomotor structures, about 20 nm across and up to 15 or 20 μm long.

Flagellation patterns are very useful in identifying bacteria and can be seen by light microscopy but only after being stained with special techniques designed to increase their thickness. The detailed structure of a flagellum can be seen only in the electron microscope.

Bacterial cells benefit from flagella in several ways. They can increase the concentration of nutrients or decrease the concentration of toxic materials near the bacterial surfaces by causing a change in the flow rate of fluids. They can also disperse flagellated organisms to areas where colony formation can take place. The main benefit of flagellated organisms is their ability to flee from areas that might be harmful.

Cell Wall

The rigid cell wall is the main structural component of most procaryotes. Some of the functions of the cell wall are (1) to provide protection for the delicate protoplast from osmotic lysis; (2) to determine a cell's shape; (3) to act as a permeability layer that excludes large molecules and various antibiotics and also plays an active role in regulating the cell's intake of ions, and (4) to provide a solid support for flagella.

The cell walls of different species may differ greatly in structure, thickness, and composition. The cell wall accounts for about 20–40% of a bacterium's dry weight.

After Christian Gram developed the Gram stain, it soon became evident that bacteria could be divided into two major groups based on their response to the Gram-stain procedure.

Gram-Positive Cell Walls

Normally, the thick, homogenous cell walls of gram-positive bacteria are composed primarily of a complex polymer, which often contains linear heteropolysaccharide chains that are bridged by peptides to form a three-dimensional netlike structure and envelop the protoplast. Gram-positive cells usually also contain large amounts of *teichoic* acids: typically, substituted polymers of ribitol phosphate and glycerol phosphate. Amino acids or sugars such as glucose are attached to the ribitol and glycerol groups.

Teichoic acids are negatively charged and help give the gram-positive cell wall its negative charge. Growth conditions can affect the composition of the cell wall; for example, the availability of phosphates affects the amount of teichoic acid in the cell wall of *Bacillus*. Teichoic acids are not present in gram-negative bacteria.

Gram-Negative Cell Walls

Gram-negative cell walls are much more complex than gram-positive walls. The gram-negative wall is about 20–30 nm thick and has a distinctly layered appearance under the electron microscope. The thin inner layer consists of peptidoglycan and constitutes no more than 10% of the wall weight. In *E. coli*, the gram-negative wall is about 1 nm thick and contains only one or two layers of peptidoglycan.

The outer membrane lies outside the thin peptidoglycan layer and is essentially a lipoprotein bilayer. The outer membrane and peptidoglycan are so firmly linked by this lipoprotein that they can be isolated as one unit.

PLASMA MEMBRANE (CYTOPLASMIC MEMBRANE)

Bordered externally by the cell wall and composed of a lipoprotein complex, the plasma membrane is the critical barrier, separating inside from outside of the cell. About 7–8 nm thick and comprising 10–20% of a bacterium's dry weight, the plasma membrane controls the passage of all material into and out of the cell. The inner and outer faces of the plasma membrane are embedded with water-loving (hydrophilic) lipids whereas the interior is hydrophobic. Control of material into the cell is accomplished by screening, as well as by electric charge. The plasma membrane is the site of the surface charge of the bacteria.

In addition to serving as an osmotic barrier that passively regulates the passage of material into and out of the cell, the plasma membrane participates in the active transport of various substances into the bacterial cell. Inside the membrane many highly reactive chemical groups guide the incoming material to the proper points for further reaction. This active transport system provides bacteria with certain advantages, including the ability to maintain a fairly constant intercellular ionic state in the presence of varying external ionic concentrations. In addition to participating in the uptake of nutrients, the cell membrane transport system participates in waste excretion and protein secretions.

CYTOPLASM

Within a cell and bounded by the cell membrane is a complicated mixture of substances and structures called the cytoplasm. The cytoplasm is a water-based fluid containing ribosomes, ions, enzymes, nutrients, storage granules (under certain circumstances), waste products, and various molecules involved in synthesis, energy metabolism, and cell maintenance.

MESOSOME

A common intracellular structure found in the bacterial cytoplasm is the mesosome. *Mesosomes* are invaginations of the plasma membrane in the shape of tubules, vesicles, or lamellae. They are seen in both gram-positive and gram-negative bacteria, although they are generally more prominent in the former.

The exact function of mesosomes is still unknown. Currently, many bacteriologists believe that mesosomes are artifacts generated during the chemical fixation of bacteria for electron microscopy (Singleton & Sainsbury, 1994).

NUCLEOID (NUCLEAR BODY OR REGION)

The nuclear region of the procaryotic cell is primitive and is a striking contrast to that of the eucaryotic cell. Procaryotic cells lack a distinct nucleus, the function of the nucleus being carried out by a single, long, double-strand of deoxyribonucleic acid (DNA) that is efficiently packaged to fit within the nucleoid. The nucleoid is attached to the plasma membrane. A cell can have more than one nucleoid when cell division occurs after the genetic material has been duplicated.

RIBOSOMES

The bacterial cytoplasm is often packed with ribosomes. Ribosomes are minute, rounded bodies, made of RNA (ribonucleic acid) and are loosely attached to the plasma membrane. Ribosomes are estimated to account for about 40% of a bacterium's dry weight; a single cell may have as many as 10,000 ribosomes. Ribosomes are the site of protein synthesis and are part of the translation apparatus.

INCLUSIONS (STORAGE GRANULES)

Storage granules or other inclusions are often seen within bacterial cells. Some inclusion bodies are not bound by a membrane and lie free in the cytoplasm. Other inclusion bodies are enclosed by a single-layered membrane about 2–4 nm thick. Many bacteria produce polymers that are stored as granules in the cytoplasm.

In procaryotic organisms, one of the most common inclusion bodies consists of poly-β-hydroxybutyric acid (PHB), which contains β-hydroxybutyrate acid units joined by ester linkage between the carboxyl and hydroxyl groups of adjacent molecules.

Another storage product is glycogen, which is a polymer of glucose units of long chains. Glycogen is dispersed evenly throughout the cytoplasm as small granules and often can be seen only with the electron microscope.

Volutin, or polyphosphate, granules are inorganic inclusion bodies that are often seen in bacterial systems. They are believed to act as reservoirs of phosphate (an important component of nucleic acids) and appear to be involved with energy metabolism. These granules show the *metachromatic effect*; that is, they appear a different shade of color than the color they were stained with.

Various sulfur-metabolizing procaryotes are capable of oxidizing and accumulating free elemental sulfur within the cell. The elemental sulfur granules remain only under conditions when excess energy nutrients are present. That is, as the sulfur is oxidized to sulfate, the granules slowly disappear.

CHEMICAL COMPOSITION

The normal growth of a bacterial cell in excess nutrients results in a cell of definite chemical composition. This growth, however, involves a coordinated increase in the mass of its constituent parts and not solely an increase in total mass.

Bacteria, in general, are composed primarily of water (about 80%) and of dry matter (about 20%). The dry matter consists of both organic (90%) and inorganic (10%) components. All basic elements for protoplasm must be derived from the liquid environment, and if the environment is deficient in vital elements, the cell will show a characteristic lack of development.

METABOLISM

Metabolism refers to the bacteria's ability to grow in any environment. The metabolic process refers to the chemical reactions that occur in living cells. In this process *anabolism* works to build up cell components, and *catabolism* breaks down or changes the cell components from one form to another.

Metabolic reactions require energy. Energy is also required for locomotion and for the uptake of nutrients. Many bacteria obtain their energy by processing chemicals from the environment through *chemosynthesis*. Other bacteria obtain their energy from sunlight through *photosynthesis*.

Chemosynthesis

The synthesis of organic substances such as food nutrients, using the energy of chemical reactions, is called chemosynthesis. A bacterium that obtains its carbon from carbon dioxide is called autotrophic. Bacteria that obtain carbon through organic compounds are called heterotrophic (see Figure 3.3).

Autotrophic Bacteria

Organisms that can synthesize organic molecules needed for growth from inorganic compounds using light or another source of energy are called *autotrophs*. For their carbon requirements, autotrophs are able to use ("fix") carbon dioxide to form complex organic compounds.

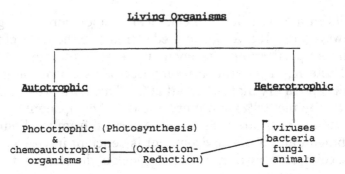

FIGURE 3.3 *Autotrophic and heterotrophic organisms in relation to their means of obtaining energy.*

Heterotrophic Bacteria

Most bacteria are not autotrophic: they cannot use carbon dioxide as a major source of carbon and therefore must rely upon the presence of more reduced, complex molecules (mostly derived from other organisms) for their carbon supply. Bacteria that need complex carbon compounds are called *heterotrophs*. The heterotrophs use a vast range of carbon sources including fatty acids, alcohols, sugars, and other organic substances. Heterotrophic bacteria are widespread in nature, and include all those species that cause disease in man, other animals, and plants.

CLASSIFICATION

Classifying microbes is not always an easy undertaking. The classification process is complicated because of the enormous variety of microorganisms that differ widely in metabolic and structural properties. Some microorganisms are plant-like, others are animal-like, and still others are totally different from all other forms of life.

As an example of the classification process, consider the microorganisms in terms of their activities: Bacteria can be classified as *aerobic, anaerobic,* or *facultative.* An aerobe must have oxygen to live. At the other extreme, the same oxygen

would be toxic to an anaerobe (lives without oxygen). Facultative bacteria are capable of growth under aerobic or anaerobic
conditions.

Because bacteria have so many forms, their proper classification or identification requires a systematic application of procedures that grow, isolate, and identify the individual varieties.
These procedures are highly specialized and technical. Ulti-
mately, the bacteria is characterized based on observation and
experience. Fortunately, as Singleton (1992) points out, certain
classification criteria have been established to help in the sort-
ing process:

1. Shape
2. Size and structure
3. Chemical activities
4. Types of nutrients they need
5. Form of energy they use
6. Physical conditions under which they can grow
7. Ability to cause disease (pathogenic or nonpathogenic)
8. Staining behavior

Using these criteria, and based on observation and experience, the bacteria can be identified from descriptions published
in *Bergey's Manual of Determinative Bacteriology* (1974).

ACTINOMYCETES

"The growth of filamentous microorganisms is the most common operational problem in the activated-sludge process"
(Metcalf & Eddy, 1991, p. 537). The fungus-like bacterial group
actinomycetes is a primary contributor to these operational
problems. Along with causing foaming in aeration basins (when
they are present in sufficient numbers and of excessive length),
actinomycetes also contribute to loss of settleability and secondary solids in wastewater treatment operations. Therefore,
gaining knowledge about this important group and studying
methodologies used in attempting to control related problems
caused by this group are of great interest to wastewater special-

ists. This section will discuss fundamental information about the group actinomycetes and especially the genus *Nocardia*.

Actinomycetes bacteria are aerobic, gram-positive organisms that form branching hyphae and asexual spores. Borne on aerial mycelia, the asexual spores are called *conidiospores* or *conidia* when they are at the tip of hyphae (see Figure 3.4) and sporangiospores when they are within the sporangia.

Although primarily soil organisms, actinomycetes have been observed in wastewater-treatment-activated sludge systems, especially in aeration and clarifier basins; they generally appear in scum layers (*Nocardia*) and, to a lesser degree, in bulking.

Actinomycetes range from rod- and cocci-shaped to filamentous organisms that, as stated previously, primarily inhabit soils where they have considerable significance. For example, they are extremely important in the mineralization of organic matter; they can degrade enormous amounts of organic compounds. Moreover, filamentous organisms can have a positive effect in wastewater treatment processes such as activated sludge. They assist in floc formation and can help to stabilize difficult wastes. However, the negative aspects of filamentous organisms in activated sludge processes receive most of the at-

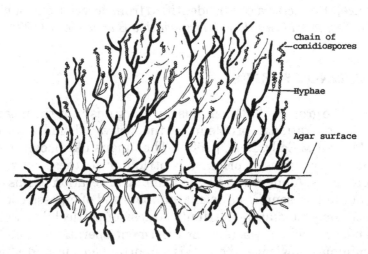

FIGURE 3.4 *Cross section of an actinomycete colony. Living hyphae appear dark and nonliving hyphae appear light. [Adapted from Prescott et al. (1993), p. 507.]*

tention and concern of wastewater specialists. For example, filamentous organisms can contribute to loss of solids to the receiving stream. Moreover, loss of solids compaction can occur. Additionally, some filamentous microorganisms such as *Microthrix parvicella*, *Nocardia* spp. and Type 1863 contribute to foaming (to be discussed in detail later).

Filamentous organisms that create the most operational problems for wastewater specialists occur primarily because of environmental and operational conditions. Conditions such as low F/M (food-to-microorganism ratio), low pH (below 6.0), septic wastewater, low nutrient level for nitrogen or phosphorus, and low dissolved oxygen can all contribute to problems. Several control measures are used to control unwanted growth of filamentous microorganisms. For example, controlling the return activated sludge (RAS) flow rate and RAS feed points; adding polymers and toxicants; treating the process to reduce impact of filamentous organisms; and adjusting pH, F/M, and dissolved oxygen; and several other corrective actions are available for use.

In the aquatic or wastewater treatment environment actinomycetes containing *Nocardia* and other filamentous microorganisms are of greatest interest and concern to wastewater operators. *Nocardia* are complex microorganisms with low pH value. They create foam and foam-related problems because they coat floc with lipid-type materials that catch air bubbles that make them buoyant. *Nocardia* are collectively called *nocardioforms*. *Nocardia*, as stated previously, are of interest to wastewater operators because they are found in and have impact on activated sludge treatment processes, especially in aeration and clarifier basins; they generally appear in scum layers (*Nocardia*) (see Figure 3.5) and, to a lesser degree, in bulking. *Nocardia* have also been identified in the biodeterioration of sewage pipes (Prescott et al., 1993).

Ideally, in the activated sludge process, the goal is to produce or develop a biomass composed mostly of bacteria. These bacteria will grow on organic materials in such a manner so as to aid gravity (by adding weight to the biomass) in settling them out in the final clarifier. The expected result is a clarified, final effluent with a thickened sludge. Unfortunately, not all treat-

(a)

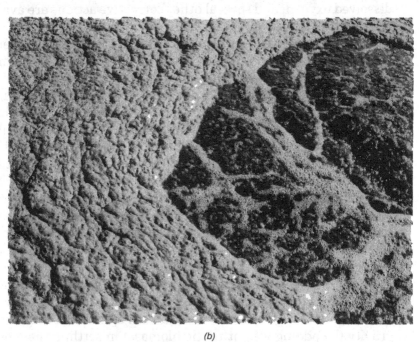

(b)

FIGURE 3.5 (a) Foaming in a large aeration basin; (b) closeup view of foaming in same basin.

34

ment processes work in a mode that produce ideal results; this less-than-ideal operation is often the case with activated sludge systems. Ideally, all bacteria in activated sludge would be of the single-cell types that grow readily in flocs. Realistically, this is not always the case because not all bacteria in activated sludge are single-cell types that grow in flocs. Many other microorganisms, such as fungi and filamentous bacteria, also may develop and cause serious operational problems: foaming and bulking.

Foaming is characterized by the formation of a viscous, brown foam that covers aeration basins and secondary clarifiers in wastewater treatment plants using the activated-sludge process. Foaming associated with wastewater treatment processes is an indicator of operational conditions or problems. For example, when a white billowy foam is present in the effluent it may indicate the presence of a high concentration of suspended solids. A light, billowy foam in the aeration basin, on the other hand, may indicate that wasting should be decreased because the sludge age is too young. When foam is dark and very thick, it usually indicates that the sludge is old and wasting should be increased (Sundstrom & Klei, 1979).

Usually, foaming is attributed to the actinomycete genus *Nocardia*. Along with causing odor problems, foaming can also lead to deterioration in effluents and can be a serious safety hazard.

Bulking occurs when slow, sludge-settling conditions are evident. Bulking is usually attributed to an excessive presence of filamentous microorganisms (Haller, 1995); typically the culprit is identified as *Nocardia*. *Nocardia* may indeed be present when bulking occurs, but it is not always the main factor, as will be pointed out later.

In order to gain a better understanding of the role that filamentous microorganisms play in foaming and bulking, it is important to take a closer look at the actinomycetes and related genera most frequently associated with these activities in the activated sludge process. The following section will provide this basic information.

Until recently, the scientific community had difficulty in classifying actinomycetes. For several years members of actinomycetes were regarded as transition forms that occupied an uncertain position with respect to the bacteria on one hand and the fungi on the other. Extensive study, however, has determined that because actinomycetes possess the general characteristics of procaryotic organisms, they should be classified as bacteria. More precisely, Singleton and Sainsbury (1994) define actinomycetes as gram-positive aerobic bacteria with species ranging from those occurring as rods to those that form (as illustrated earlier) both substrate mycelium and/or aerial mycelium. These bacteria are saprophytes (obtain nutrients from nonliving and decaying plant or animal matter) that can grow on a wide range of organic compounds such as long-chained hydrocarbons, dead microbial mass, complex aromatics, and wastes that are difficult to degrade. They can also grow on (but at a slower rate) amino acids, sugars, and wastes with low molecular weight. Once an aggressive population of actinomycetes has been established, it readily utilizes wastes in large quantities. The simplest actinomycete forms can attain maximum growth in about 24 hours and in about 8–20 days for more advanced forms. Some actinomycete forms have been identified as pathogenic (disease causing). The pathogenic forms include *Mycobacterium* spp., *Nocardia asteriodes*, and *Actinomyces israelii*.

NOCARDIA

The term *Nocardia* is not foreign or unknown to wastewater operators. Indeed, when foaming conditions appear on an aeration basin in a wastewater treatment plant, it is not unusual for the plant operator to assume automatically that *Nocardia* is the cause. In some cases this might be the correct assumption, in others it might not. While it is true that biological foams can be induced by actinomycete microorganisms, such as *Nocardia*, it is also true that foaming can be caused by nutrient deficiency and other filamentous bacteria such as *Microthrix parvicella*. If the foam is actinomycetes induced, several different related genera besides *Nocardia* may be present. For example, *Mycobacterium*, *Micromonospora*, *Arthrobacter*, *Actinomadura*, and *Corynebacterium* may be present.

When *Nocardia* is the primary genera present in the activated sludge process (especially in aeration basins), it may be of several different species: *N. amarae, N. caviae, N. pinesis,* and/or *N. rhodochrus.*

Excessive *Nocardia* presence in the activated sludge process usually forms a characteristic foam that is viscous, stable, and often chocolate brown in color. *Nocardia,* as stated previously, is of concern to wastewater specialists not only because of foam and foam-related problems but also because of problems related to the loss of secondary solids and the creation of safety problems. Safety problems may go beyond the hazards associated with slipping and sliding on carried-over foam. As a case in point, consider the results of the study of actinomycetes in activated sludge systems conducted by Lechevalier et al. (1977). Their findings pointed out that several isolated species of *Nocardia* and *Mycobacterium* were identified as human pathogens.

At this point it should be pointed out that not all aspects of *Nocardia* growth in wastewater treatment processes are negative. That is, *Nocardia* can also play a beneficial role by feeding on a large variety of complex wastes. Moreover, *Nocardia* can also assist with floc formation and, thus, with clarification.

Along with foaming problems, wastewater treatment specialists must also deal with floating sludge or sludge bulking problems. *Nocardia* plays a role in floating sludge or sludge bulking problems but is not the sole contributor (as some would suggest). This is the case because actinomycetes (and *Nocardia*) are short in length and generally present mostly within floc particles rather than extending from the surface of floc particles. Thus, the production of floating sludge or sludge bulking is caused by other factors that might be working in conjunction with numerous nocardial filaments. These other factors might include the presence of insoluble lipids and entrained gases that contribute to floating sludge or bulking sludge. An important point to keep in mind is that when floating-sludge or bulking-sludge problems are actinomycete-related, floating sludge normally appears in secondary clarifiers. Moreover, foaming and floating sludge problems can be carried over to anaerobic digesters, thickeners, and solids dewatering facilities.

In addition to safety and the other problems discussed earlier, foam carryover can lead to additional waste-stream problems, such as loss of secondary solids to the receiving stream; an increase in sludge handling costs; and decreased thickener, anaerobic digester, and dewatering facilities efficiencies.

Why Do Actinomycetous Microorganisms End up in the Wastewater Treatment Process?

Several operational and environmental factors have been attributed to favoring the growth of actinomycete-related microorganisms. For example, if the pH level is less than 6.5 and grease and oil loading are high, environmental conditions may be suitable for actinomycete growth. In addition, when low aeration basin dissolved oxygen concentration and low organic loading rate (low F/M) occur, conditions are favorable for growth of filamentous organisms found in activated sludge (Strom & Jenkins, 1984). In regard to *Nocardia* spp. (specifically as an agent being responsible for foaming) (Pipes, 1978; Dhaliwal, 1979) determined that the key factors associated with *Nocardia*-induced foaming include sludge ages greater than 8 days, high oil and grease content, and warmer temperatures.

Because several potential causal factors have been identified as elements that lead to foaming and/or bulking problems in activated sludge systems, it follows that several different strategies have been developed and attempted to control actinomycete growth. Suggested correction measures include (1) building baffles across chlorine contact basins, (2) using bioaugmentation products and defoaming agents, and (3) decreasing the aeration rate in aeration basins. Other corrective practices include using cationic polymers, which work to collapse the foam, using sodium hypochlorous solution to control foam formation and seed recycling, which is accomplished by treating returning supernatants and solids to the activated sludge process for the presence of actinomycetes. Treating is accomplished by applying chlorine. Other measures that have been attempted to control foaming in activated sludge include chlorinating RAS, changing the mode of operation, controlling hydrocarbon wastes, raising pH levels without using lime, reducing mixed-liquor suspended solids (MLSS), and using selectors. Selectors are separate or sectionalized compartments that are the initial contact zone of a biological reactor; that is, a selector is the place where the primary effluent and return activated sludge are mixed or combined. The selector provides an environment where rapid adsorption of soluble organics into floc-forming microorganisms occurs; thus, little soluble organic material is left for the filamentous microorganisms (Metcalf & Eddy, 1991). As a final resort (actually this technique is used quite often, unfortunately), biological foams must be physically removed from the wastewater treatment process(es). This is a time-consuming, labor-intensive, and dangerous operation. For those who may have experienced the thrill of sliding (unexpectedly) on wet polymers, sliding on aeration basin foam can be just as dangerous. Have you ever experienced the thrill of manually removing frozen foam from a treatment process? Hopefully not. The point to keep in mind is that even when foam is not frozen, it can still have the "consistency of styrofoam" and be difficult to remove (Tchobanoglous & Schroeder, 1987, p. 619).

Research has been conducted to identify the types of filamentous organisms found in activated sludge. In addition, much of this research has attempted to relate the occurrence of filamentous organisms to activated sludge operational problems, that is, to problems associated with bulking and foaming, and to determine corrective measures to prevent the growth of unwanted species.

Probably the most definitive and best known research effort (to date) was conducted by Michael G. Richard, Ph.D., of the Department of Microbiology and Environmental Health, Colorado State University. Dr. Richard's research focused on filamentous microorganisms and activated sludge operations at several wastewater treatment facilities in Colorado. Specifically, he set out to identify the operational problems associated with filamentous microorganisms in activated sludge processes and to diagnose and recommend remedial actions.

The results of Dr. Richard's research were presented in 1986 at the Annual Meeting of the Rocky Mountain Section of the

American Water Works Association/Water Pollution Control Association, Breckenridge, Colorado. A brief summary of his findings follows.

Dr. Richard's findings indicated that foaming and bulking problems occurred in several of Colorado's activated sludge plants. It is interesting to note that in every case of sludge bulking, sufficient amounts of filamentous organisms were observed and accounted for the bulking. However, this was not true in every case of foaming; approximately 80% of foaming episodes could be attributed to filamentous growth. In the foams examined, *Nocardia* spp. and *Microthrix parvicella* were always present in large concentrations. Moreover, the findings indicated seasonal variations in foaming and in bulking occurrences. Several occurrences of bulking and foaming were operationally related and linked to system operational efficiency. Cases where carried-over foam froze and then had to be manually removed were reported. Several attempts were made to control bulking and foaming. The findings indicated that many of these control measures were unsuccessful. The main result of Dr. Richard's study was the finding that further remedial control methods are needed and more research on filamentous microorganisms should be done (Richard, 1986).

As has been made clear in this section, the actinomycetes group of microorganisms play an active role in wastewater treatment and operational problems. For those wastewater specialists who have a fundamental knowledge of biology and microbiology and wish to pursue studies beyond the fundamental aspects of foaming and bulking problems in wastewater treatment, Wanner (1994) discusses these topics in great detail and in easy to understand terms in his text *Activated Sludge Bulking and Foaming Control*.

VIRUSES

This chapter ends with a discussion of viruses. Viruses are included here because they (bacteriophages) have a relationship to bacteria and for lack of a better place to include these important agents of disease.

Viruses are parasitic particles that are the smallest living infectious agents known. They are not cellular in that they have no nucleus, cell membrane, or cell wall. They multiply only within living cells (hosts) and are totally inert outside of living cells but can survive in the environment. It takes only a single virus cell to infect a host. As far as measurable size goes, viruses range from 20–200 millimicrons in diameter, about one to two orders of magnitude smaller than bacteria. Humans excrete more than 100 virus types through the enteric tract that can find their way into sources of drinking water. In sewage the average is between 100–500 enteric infectious units/100 ml. If the viruses are not killed in the treatment process and become diluted by the receiving stream to 0.1–1 viral infectious units/100 ml, the low concentrations make it very difficult to determine virus levels in water supplies. Since tests are usually run on samples of less than 1 ml, at least 1,000 samples would have to be analyzed to detect a single virus unit in a liter of water. For this reason, the viruses are usually concentrated by filtration or centrifugation prior to analysis (Sundstrom & Klei, 1979).

Viruses differ from living cells in at least three ways: (1) they are unable to reproduce independently of cells and carry out cell division, (2) they possess only one type of nucleic acid, either DNA or RNA, and (3) they have a simple acellular organization (Prescott et al., 1993). Viruses can be controlled by chlorination but at much higher levels than are necessary to kill the bacteria. Some viruses that may be transmitted by water include hepatitis A, adenovirus, polio, coxsackie, echo, and Norwalk agent. A virus that infects a bacterium is called a *bacteriophage*.

Bacteriophage

Lewis Thomas (1974) pointed out that when humans "catch diphtheria it is a virus infection, but not of us." That is, when humans are infected by the virus causing diphtheria, it is the bacterium that is really infected—humans simply "blundered into someone else's accident" (p. 76). The toxin of diphtheria bacilli is produced when the organism has been infected by bacteriophage.

A bacteriophage (phage) is any viral organism whose host is a bacterium. Most bacteriophage research has been carried out on the bacterium *Escherichia coli*, which is one of the gram-negative bacteria that water and wastewater specialists are concerned about because it is a typical coliform (to be discussed later).

As stated earlier, a virus does not have a cell-type structure from which it is able to metabolize or reproduce. However, when the *genome* (a complete haploid set of chromosomes) of a virus is able to enter a viable living cell (a bacterium), it may "take charge" and direct the operation of the cell's internal processes. When this occurs, the genome, through the host's synthesizing process, is able to reproduce copies of itself, move on, and then infect other hosts. Hosts of a phage may involve a single bacterial species or several bacteria genera.

The most important properties used in classifying bacteriophages are nucleic acid properties and phage morphology.

Bacterial viruses may contain either DNA or RNA; most phages have double-stranded DNA.

Many basic structures have been recognized among phages. Phages appear to show greater variation in form than any other viral group. The basic morphological structure of the T-2 bacteriophage is illustrated in Figure 3.6. From Figure 3.6 it is apparent that the T-2 phage has two prominent structural characteristics: the head (a polyhedral capsid) and the tail.

The effect of phage infection depends on the phage and host and to a lesser extent on conditions. Some phages multiply

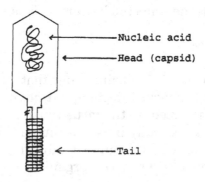

FIGURE 3.6 *A simplified diagram shows the major components of a T-2 phage.*

within and lyse (destroy) their hosts (Singleton, 1992). When the host lyses (dies and breaks open), phage progeny are released.

SUMMARY OF KEY TERMS

- *Actinomycete:* an aerobic, gram-positive bacterium that forms branching hyphae (filaments) and is commonly associated with wastewater treatment problems such as bulking and foaming
- *Bacteria:* procaryotes that occur almost everywhere. They may be spherical (cocci), rod shaped (bacilli), or spiral. Other forms of bacteria may resemble fungi or have no characteristic shape at all. In most cases a bacterium is a single, autonomous cell; however, they may remain together after division to form pairs, chains, and clusters of various shapes and sizes.
- *Bacteriophage:* a virus that infects bacteria; often called a phage
- *Bulking:* a wastewater treatment operational problem that is identified when sludge settles but does not compact quickly. Bulking is sometimes caused by filamentous microorganisms (Haller, 1995).
- *Gram stain:* a differential staining procedure by which bacteria are categorized as either gram-positive or gram-negative based on their ability to retain a primary dye when decolorized with an organic solvent such as ethanol
- *Foaming:* a wastewater treatment operational condition that can be caused by filamentous microorganisms
- *Virus:* an infectious agent having a simple acellular organization with a protein coat and a single type of nucleic acid and reproducing only within living host cells

CHAPTER 4

Fungi

THE fungi (singular fungus) constitute an extremely important and interesting group of eucaryotic, aerobic microbes ranging from the unicellular yeasts to the extensively mycelial molds. Not considered plants, they are a distinctive life form of great practical and ecological importance. Fungi are important because, like bacteria, they metabolize dissolved organic matter; they are the principal organisms responsible for the decomposition of carbon in the biosphere. Fungi, unlike bacteria, can grow in low moisture areas and in low pH solutions, which aids in the breakdown of organic matter.

Fungi comprise a large group of organisms that include such diverse forms as the water molds, slime molds, other molds, mushrooms, puffballs, and yeasts. Because they lack chlorophyll (and thus are not considered plants), they must get nutrition from organic substances. They are either parasites, existing in or on animals or plants, or more commonly are *saprophytes*, obtaining their food from dead organic matter. The fungi belong to the kingdom *Myceteae*. The study of fungi is called *mycology*.

It is interesting to note that McKinney (1962) in his well-known text, *Microbiology for Sanitary Engineers*, complains that the study of mycology has been directed solely toward classification of fungi and not toward the actual biochemistry involved with fungi. McKinney points out that it is important for those involved in the sanitation field to recognize the "sanitary importance of fungi . . . and other steps will follow" (p. 40). The water and wastewater operator needs to understand the role of

45

fungi as it relates to the water purification process. Moreover, wastewater specialists need knowledge and understanding of the organism's ability to function and exist under extreme conditions, which make them important elements in biological wastestream treatment processes, and the degradation that takes place during composting.

Fungi may be unicellular or filamentous. They are large, 5–10 microns wide, and can be identified by a microscope. The distinguishing characteristics of the group, as a whole, include the following: (1) nonphotosynthetic, (2) lack tissue differentiation, (3) have cell walls of polysaccharides (chitin), and (4) propagate by spores (sexual or asexual).

CLASSIFICATION

The fungi are divided into five classes:

1. Myxomycetes, or slime fungi
2. Phycomycetes, or aquatic fungi (algae)
3. Ascomycetes, or sac fungi
4. Basidiomycetes, or rusts, smuts, and mushrooms
5. Fungi imperfecti, or miscellaneous fungi

Although fungi are limited to only five classes, there are more than 80,000 known species. With respect to water quality, the first two classes are of greatest importance (Tchobanglous & Schroeder, 1987).

IDENTIFICATION

Fungi differ from bacteria in several ways including their size, structural development, methods of reproduction, and cellular organization. The structure of fungi, not their biochemical reactions (unlike the bacteria), is used to identify them. Since fungi can be examined directly or suspended in liquid, stained, and dried, they can be observed under microscopic examination. Under the microscope, for example, several fungi can be identified by the appearance (color, texture, and diffusion of pigment) of their mycelia. The experienced water or

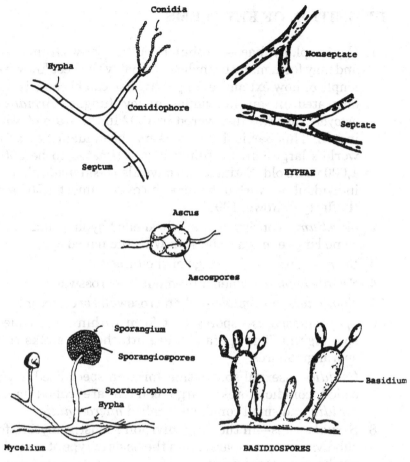

FIGURE 4.1 *Nomenclature of fungi. [Adapted from McKinney (1962), p. 36.]*

wastewater specialist usually has little difficulty in identifying typical fungal species.

One of the tools available to the water or wastewater specialist for use in the fungal identification process is the distinctive terminology used in mycology. Fungi go through several phases of their life cycle and their structural characteristics change with each phase. Therefore, it is important to become familiar with the terms that are listed and defined in the following section. As a further aid in learning how to identify fungi, it is helpful to relate the defined terms to their diagrammatic representations as shown in Figure 4.1.

DEFINITION OF KEY TERMS

1. *Hypha* (pl. hyphae)—a tubular cell that grows from the tip and may form many branches. Probably the best known example of how extensive fungal hyphae can become is demonstrated in an individual honey fungus, *Armalloria ostoyae*, that was discovered in 1992 in the state of Washington. This particular fungus has been identified as the world's largest living thing; it is estimated to be 500 to 1,000 years old. Estimations have also been made about its individual network of hyphae: it covers almost 1,500 acres (Lafferty & Rowe, 1993).

2. *Mycelium*—consists of many branched hyphae and can become large enough to be seen with the naked eye

3. *Spore*—reproductive stage of the fungi

4. *Septate hyphae*—when a filament has crosswalls

5. *Nonseptate* or *aseptate*—when crosswalls are present

6. *Sporangiospores*—spores that form within a sac called a sporangium. The sporangia are attached to stalks called sporangiophores.

7. *Conidia*—asexual spores that form on specialized hyphae called conidiophores. Large conidia are called *macroconidia* and small conidia are called *microconidia*.

8. *Sexual spores*—in the fungi division Amastigomycota four subdivisions are separated on the basis of type of sexual reproductive spores present.
   - subdivision Zygomycotina—consists of nonseptate hyphae and zygospores. Zygospores are formed by the union of nuclear material from the hyphae of two different stains.
   - subdivision Ascomycotina—fungi in this group are commonly referred to as the ascomycetes. They are also called sac fungi. They all have septate hyphae. Ascospores are the characteristic sexual reproductive spores and are produced in sacs called asci (ascus, singular). The mildews and *Penicillium* with asci in long fruiting bodies belong to this group.
   - subdivision Basidiomycotina—consists of mushrooms, puffballs, smuts, rust, and shelf fungi that are found on

dead trees. The sexual spores of this class are known as basidiospores, which are produced on the club-shaped basidia.

- subdivision Deutermycotina—consists of only one group, the Deuteromycetes. Members of this class are referred to as the fungi imperfecti and include all the fungi that lack sexual means of reproduction.

9. *Budding*—process by which yeasts reproduce
10. *Blastospore* or bud—spores formed by budding

CULTIVATION OF FUNGI

Fungi can be grown and studied by culturing methods. When culturing fungi, it is important to use culture media that limit the growth of other microbial types. For example, controlling bacterial growth is of particular importance. This can be accomplished by using special agar that depresses pH of the culture medium (usually Sabouraud glucose or maltose agar) to prevent the growth of bacteria. In addition, antibiotics can be added to the agar to prevent bacterial growth.

REPRODUCTION

As part of their reproductive cycle, fungi produce very small spores that are easily suspended in air and widely dispersed by the wind. Fungal spores are also spread by insects and other animals. The color, shape, and size of spores are useful in the identification of fungal species.

Reproduction in fungi can be either sexual or asexual. Sexual reproduction is accomplished by the union of compatible nuclei. Specialized asexual and/or sexual spore-bearing structures (fruiting bodies) are formed by most fungi. Some fungal species are self-fertilizing and other species require outcrossing between different but compatible vegetative thalluses (mycelia).

Most fungi are asexual. Asexual spores are often brightly pigmented, giving the colony a characteristic color (e.g., green, red, brown, black, blue—the blue spores of *Penicillium roquefort* are found in blue or Roquefort cheese).

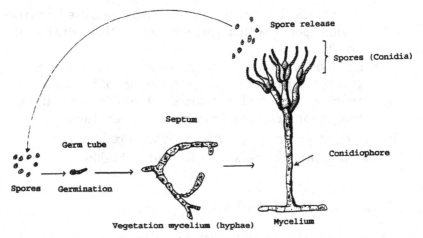

FIGURE 4.2 *Asexual life cycle of Penicillium sp. [Adapted from Wistreich and Lechtman (1980), p. 163.]*

Asexual reproduction is accomplished in several ways:

1. Vegetative cells may *bud* to produce new organisms. This is very common in the yeasts.
2. A parent cell can divide into two daughter cells.
3. The most common method of asexual reproduction is the production of spores (see Figure 4.2). There are several types of asexual spores:
   - If the hypha separates to form cells (arthrospores), they behave as spores.
   - If the cells are enclosed by a thick wall before separation, they are called chlamydospores.
   - If the spores are produced by budding, they are called blastospores.
   - If the spores develop within a sporangia (sac), they are called sporangiospores.
   - If the spores are produced at the sides or tips of the hypha, they are called conidiospores.

NUTRITION AND METABOLISM

Fungi are found wherever organic material is available. They prefer moist habitats and grow best in the dark. Most fungi are saprophytes, acquiring their nutrients from dead organic matter. This type of nutrition is gained when the fungi secrete hydrolytic enzymes, which digest external substrates; they are able to use dead organic matter as a source of carbon and energy. Most fungi use glucose and maltose (carbohydrates) and nitrogenous compounds to synthesize their own proteins and other needed materials. Knowing from what materials fungi synthesize their own protein and other needed materials in comparison to what bacteria are able to synthesize helps the water/wastewater operator understand the growth requirements of the different microorganisms (McKinney, 1962).

SUMMARY OF KEY TERMS

- *Fungi:* unicellular or multicellular, eucaryotic, heterotrophic microorganisms that do not contain chlorophyll and typically form a rigid cell wall (Singleton & Sainsbury, 1994)
- *Saprotroph:* an organism that feeds on waste or dead organic matter

CHAPTER 5

Algae

ONE does not have to be a water or wastewater specialist to understand that algae can be a nuisance. This is the case because many ponds, lakes, rivers and streams in the United States are currently undergoing *eutrophication*, which is the enrichment of an environment with inorganic substances (e.g., phosphorus and nitrogen). The average person may not know what eutrophication means. However, when eutrophication occurs and when filamentous algae like *Caldophora* break loose in a pond, lake, stream, or river and wash ashore near where this person is standing, algae makes its stinking, noxious presence known.

For the water or wastewater specialist, algae are both a nuisance and a valuable ally. Although they are not pathogenic, algae do cause problems with water treatment operations. They grow easily on the walls of troughs and basins, and heavy growth can cause plugging of intakes and screens. Additionally, algae release chemicals that often give off undesirable tastes and odors. In wastewater treatment, on the other hand, controlled algae growth can be valuable in long-term oxidation ponds where they aid in the purification process by producing oxygen.

Before beginning a detailed discussion of algae, key terms are defined.

DEFINITION OF KEY TERMS

1. *Algae*—refers to a large and diverse assemblage of eucaryotic organisms that lack roots, stems, and leaves but

have chlorophyll and other pigments for carrying out oxygen-producing photosynthesis.

2. *Algology* or *phycology*—the study of algae
3. *Antheridium*—special male reproductive structure where sperm are produced
4. *Aplanospore*—nonmotile spores produced by sporangia
5. *Benthic*—algae attached and living on the bottom of water body
6. *Binary fission*—nuclear division followed by division of the cytoplasm
7. *Chloroplasts*—packets that contain chlorophyll a and other pigments
8. *Chrysolaminarin*—the carbohydrate reserve in organisms of division *Chrysophyta*
9. *Diatoms*—photosynthetic, circular, or oblong chrysophyte cells
10. *Dinoflagellates*—unicellular, photosynthetic protistan algae
11. *Epitheca*—the larger part of the frustule (diatoms)
12. *Euglenoids*—contain chlorophylls a and b in their chloroplasts; representative genus is *Euglena*
13. *Fragmentation*—a type of asexual algal reproduction in which the thallus breaks up and each fragmented part grows to form a new thallus
14. *Frustule*—the distinctive two-piece wall of silica in diatoms
15. *Hypotheca*—the small part of the frustule (diatoms)
16. *Neustonic*—algae that live at the water-atmosphere interface
17. *Oogonia*—vegetative cells that function as female sexual structures in algal reproductive system
18. *Pellicle*—a *Euglena* structure that allows for turning and flexing of the cell
19. *Phytoplankton*—made up of algae and small plants
20. *Plankton*—consists of free-floating, mostly microscopic aquatic organisms

21. *Planktonic*—refers to algae that are suspended in water as opposed to attached and living on the bottom (benthic)

22. *Protothecosis*—a disease in humans and animals that is caused by the green algae, *Prototheca moriformis*

23. *Thallus*—the vegetative body of algae

ALGAE—DESCRIPTION

Algae are autotrophic, contain the green pigment chlorophyll, and are a form of aquatic plants. Algae differ from bacteria and fungi in their ability to carry out photosynthesis—the biochemical process requiring sunlight, carbon dioxide, and raw mineral nutrients. Photosynthesis takes place in the chloroplasts. The chloroplasts are usually distinct and visible. They vary in size, shape, distribution, and number. In some algal types the chloroplast may occupy most of the cell space. Chloroplasts usually grow near the surface of water because light cannot penetrate very far through water. Although in mass they are easily seen by the unaided eye (multicellular forms like marine kelp), many chloroplasts are microscopic. Algal cells may be motile by one or more flagella, exhibit gliding motility as in diatoms, or nonmotile. They occur most commonly in water (fresh and polluted water, as well as in salt water) in which they may be suspended (planktonic) phytoplanktons or attached and living on the bottom (benthic). A few algae live at the water-atmosphere interface and are termed neustonic. Within the fresh and saltwater environments, algae are important primary producers (the start of the food chain for other organisms). During their growth phase, they are important oxygen-generating organisms and constitute a significant portion of the plankton in water.

CHARACTERISTICS USED IN CLASSIFYING ALGAE

According to the five-kingdom system of Whittaker, the algae belong to seven divisions distributed in two different kingdoms. Although there are seven divisions of algae, only five divisions are of primary interest to water and wastewater specialists and are listed as follows:

- *Chlorophyta*—green algae
- *Euglenophyta*—euglenoids
- *Chrysophyta*—golden-brown algae, diatoms
- *Phaeophyta*—brown algae
- *Pyrrophyta*—dinoflagellates

The primary classification of algae is based on cellular properties. Several characteristics are used to classify algae including (1) cellular organization and cell wall structure; (2) the nature of the chlorophyll(s); (3) the type of motility, if any; (4) the carbon polymers that are produced and stored; and (5) the reproductive structures and methods. Table 5.1 summarizes the properties of the five divisions discussed in this text.

Algal Cell Wall

Algae show considerable diversity in the chemistry and structure of their cell walls. For example, their cell wall is a thin, rigid structure usually composed of cellulose that is usually modified by the addition of other polysaccharides. In other algae, the cell is sometimes strengthened by the deposition of calcium carbonate. Other forms have chitin present in the cell wall. Complicating the classification of algal organisms are the *euglenoids*, which lack cell walls. In diatoms the cell wall is composed of silica. The frustules of diatoms have extreme resistance to decay and remain intact for long periods of time, as the fossil records indicate.

Chlorophyll(s)

The principal feature used to distinguish algae from other microorganisms (e.g., fungi) is the presence of chlorophyll and other photosynthetic pigments in the algae. All algae contain chlorophyll a. Some, however, contain other types of chlorophyll. The presence of these additional chlorophylls is characteristic of a particular algal group. In addition to chlorophyll, other pigments encountered in algae include fucoxanthin (brown), xanthophylls (yellow), carotenes (orange), phycocyanin (blue), and phycoerythrin (red).

Table 5.1. Comparative summary of algal characteristics.

Algal Group	Common Name	Structure	Pigments	Carbon Reserve Materials	Motility	Method of Reproduction
Chlorophyta	Green algae	Unicellular to multicellular	Chlorophylls a and b, carotenes, xanthophylls	Starch, oils	Most are nonmotile	Asexual and sexual
Euglenophyta	Euglenoids	Unicellular	Chlorophylls a and b, carotenes, xanthophylls	Fats	Motile	Asexual
Chrysophyta	Golden-brown algae, diatoms	Unicellular	Chlorophylls a and b, special carotenoids, xanthophylls	Oils	Gliding by diatoms; others by flagella	Asexual and sexual
Phaeophyta	Brown algae	Multicellular	Chlorophylls a and b, carotenoids, xanthophylls	Fats	Motile	Asexual and sexual
Pyrrophyta	Dinoflagellates	Unicellular	Chlorophylls a and b, carotenes, xanthophylls	Starch, oils	Motile	Asexual; sexual rare

Motility

Many algae have flagella (threadlike appendages). The flagella are locomotor organelles that may be the single polar or multiple polar types. The *Euglena* is a simple flagellate form with a single polar flagellum. Chlorophyta, on the other hand, have either two or four polar flagella. Dinoflagellates have two flagella of different lengths. In some cases, algae are nonmotile until they form motile gametes (a haploid cell or nucleus) during sexual reproduction. Diatoms do not have flagella but have gliding motility.

Algal Nutrition

Algae can be either autotrophic or heterotrophic. Most are photoautotrophic; they require only carbon dioxide and light as their principal sources of energy and carbon. In the presence of light, algae carry out oxygen-evolving photosynthesis; in the absence of light, algae utilize oxygen. Chlorophyll and other pigments are used to absorb light energy for photosynthetic cell maintenance and reproduction. One of the key characteristics used in the classification of algal groups is the nature of the *reserve polymer* synthesized as a result of utilizing carbon dioxide present in water.

Algal Reproduction

Algae may reproduce either asexually or sexually. There are three types of asexual reproduction: binary fission, spores, and fragmentation. In some unicellular algae, *binary fission* occurs where nuclear division is followed by the division of the cytoplasm, forming new individuals like the parent cell. Some algae reproduce through *spores*. These spores are unicellular and germinate without fusing with other cells. In *fragmentation*, the thallus breaks up and each fragment grows to form a new thallus.

Sexual reproduction can involve union of cells where eggs are formed within vegetative cells called *oogonia* (function as female structures) and sperm are produced in a male reproductive organ called an *antheridium*. Algal reproduction can also occur through a reduction of chromosome number and/or the union of nuclei.

CHARACTERISTICS OF ALGAL DIVISIONS

Chlorophyta (Green Algae)

The majority of algae found in ponds belong to this group; they also can be found in salt water and the soil. Several thousand species of green algae are known today. Many are unicellular, while some are multicellular filaments or aggregated colonies. The green algae have chlorophylls a and b along with specific carotenoids, and they store carbohydrates as starch. Few green algae are found at depths greater than 7–10 meters, largely because sunlight does not penetrate to that depth. Some species have a holdfast structure that anchors them to the bottom of the pond and to other submerged inanimate objects. Green algae reproduce by both sexual and asexual means.

Euglenophyta (Euglenoids)

Euglenoids are a small group of unicellular microorganisms that have a combination of animal and plant properties. For example, euglenoids lack a cell wall, possess a gullet, have the ability to ingest food, have the ability to assimilate organic substances, and, in some species, are absent of chloroplasts. They occur in fresh, brackish, and salt waters and on moist soils. A typical *Euglena* cell is elongated and bounded by a plasma membrane; the absence of a cell wall makes it very flexible in movement. Inside the plasma membrane is a structure called the *pellicle*, which gives the organism a definite form and allows for turning and flexing of the cell. Euglenoids that are photosynthetic contain chlorophylls a and b, and they always have a red eyespot (*stigma*), which is sensitive to light. Some euglenoids move about by means of flagellum; others move about by means of contracting expanding motions. The characteristic food supply for euglenoids is a lipopolysaccharide. Reproduction in euglenoids is by simple cell division.

Chrysophyta (Golden-Brown Algae)

The Chrysophycophyta group is quite large and diversified with several thousand members. They differ from green algae and euglenoids in that (1) chlorophylls a and c are present, (2) fucoxanthin, a brownish pigment, is present, and (3) they store food in the form of oils and leucosin, a polysaccharide. The combination of yellow pigments, fucoxanthin, and chlorophylls causes most of these algae to appear golden brown. The Chrysophycophyta group is also diversified in cell wall chemistry and flagellation and has three major classes: golden-brown algae, yellow-brown algae, and diatoms.

Some Chrysophyta lack cell walls; others have intricately patterned coverings external to the plasma membrane, such as walls, plates, and scales. The diatoms are unique in that they have hard cell walls of pectin, cellulose, or silicon that are constructed in two halves (the *epitheca* and the *hypotheca*), called a *frustule*. Two anteriorly attached flagella are common among Chrysophyta; others have no flagella.

Most Chrysophyta are unicellular or colonial. Asexual cell division is the usual method of reproduction in diatoms; other forms of Chrysophyta can reproduce sexually.

Diatoms have direct and indirect economic significance for humans. Because they make up most of the phytoplankton of the cooler ocean parts, they are the ultimate source of food for fish. Water and wastewater operators understand the ability of diatoms to function as indicators of industrial water pollution. As water quality indicators, their specific tolerances to environmental parameters such as pH, nutrients, nitrogen, concentration of salts, and temperature, have been compiled (Prescott et al., 1993).

Phaeophyta (Brown Algae)

With the exception of a few freshwater species, all algal species of this division exist in marine environments as seaweeds.

They are a highly specialized group, consisting of multicellular organisms that are sessile (i.e., attached and not free moving). These algae contain essentially the same pigments seen in the golden-brown algae, but they appear brown because of the predominance of and the masking effect of a greater amount of fucoxanthin. Brown algal cells store food as the carbohydrate laminarin and some lipids. The brown algae reproduce asexually.

Rhodophyta (Dinoflagellates)

The principal members of this division are the *dinoflagellates*. The dinoflagellates comprise a diverse group of biflagellated and nonflagellated unicellular, eucaryotic organisms. The dinoflagellates occupy a variety of aquatic environments with the majority living in marine habitats. Most of these organisms have a heavy cell wall composed of cellulose-containing plates. They store food as starch, fats, and oils. These algae have chlorophylls a and c and several xanthophylls. The most common form of reproduction in dinoflagellates is by cell division, but sexual reproduction has also been observed.

SUMMARY OF KEY TERMS

- *Algae:* eucaryotic plants that lack roots, stems, and leaves but have chlorophyll and other pigments for carrying out photosynthesis

CHAPTER 6

Protozoa and Other Microorganisms

PROTOZOA

THE protozoa ("first animals") are a large group of eucaryotic organisms of more than 50,000 known species that have adapted a form or cell to serve as the entire body. In fact, all protozoans are single-celled organisms. Typically, they lack cell walls but have a plasma membrane that is used to take in food and discharge waste. They can exist as solitary or independent organisms, for example, the stalked ciliates such as *Vorticella* sp., or they can colonize like the sedentary *Carchesium* sp. Protozoa are microscopic and get their name because they employ the same type of feeding strategy as animals. Most are harmless, but some are parasitic. Some forms have two life stages: active trophozoites (capable of feeding) and dormant cysts.

As unicellular eucaryotes, protozoa cannot be easily defined because they are diverse and, in most cases, only distantly related to each other (Patterson & Hedley, 1992). Each protozoan is a complete organism and contains the facilities for performing all the body functions for which vertebrates have many organ systems.

Pathogenic Protozoa

Certain types of protozoans can cause disease. Of particular interest to the drinking water practitioner are the *Entamoeba histolytica* (amoebic dysentery and amoebic hepatitis), *Giardia lamblia* (giardiasis), and *Cryptosporidium* (cryptosporidiosis).

Sewage contamination transports eggs, cysts, and oocysts of parasitic protozoa and helminths (tapeworms, hookworms, etc.) into raw water supplies, leaving it to water treatment and disinfection to diminish the danger of contaminated water to the consumer.

To prevent the occurrence of *Giardia* and *Cryptosporidium* spp. in surface water supplies, and to address the increasing problem of waterborne diseases, the U.S. EPA implemented its Surface Water Treatment Rule (SWTR) on 29 June 1989. The rule requires both filtration and disinfection of all surface water supplies as the primary means of controlling *Giardia* spp. and enteric virus. Since implementation of its Surface Water Treatment Rule, U.S. EPA has also recognized that *Cryptosporidium* spp. is an agent of waterborne disease (Badenock, 1990). In its next series of surface water regulations (1996), the U.S. EPA included *Cryptosporidium*.

To test the need for and the effectiveness of U.S. EPA's Surface Water Treatment Rule, LeChevallier et al. (1991) conducted a study on the occurrence and distribution of *Giardia* and *Cryptosporidium* organisms in raw water supplies to 66 surface water filter plants. These water filter plants were located in 14 states and one Canadian province. A combined immunofluorescence test indicated that cysts and oocysts were widely dispersed in the aquatic environment. *Giardia* spp. were detected in >80% of the samples. *Cryptosporidium* spp. were found in 85% of the sample locations. Taking into account several variables, *Giardia* or *Cryptosporidium* spp. were detected in 97% of the raw water samples. After evaluating their data, the researchers came to the conclusion that the Surface Water Treatment Rule may have to be upgraded (and subsequently has been) to require additional treatment.

In 1998, U.S. EPA finalized the long-awaited Stage 1 Disinfectants/Disinfection By-products (D/DBP) and Interim Enhanced Surface Water Treatment rules (IESWT Rule) that tighten controls on DBPs and turbidity, and regulate *Cryptosporidium* for the first time. Highlights of the IESWT Rule follow. (Note: Long-term control strategies are expected to be implemented after the year 2000.)

IESWT Rule: This treatment optimization rule, which only

applies to large (those serving more than 10,000 people) public water systems that use surface water or groundwater directly influenced by surface water, is the first to directly regulate *Cryptosporidium* (crypto).

The rule

- sets a crypto maximum contaminant level goal (MCLG) of zero
- requires systems that filter to remove 99% (2 log) of crypto oocysts
- adds crypto control to watershed protection requirements for systems operating under filtration waivers
- is particular to the genus *Cryptosporidium*, not to the *Cryptosporidium parvum* species

Giardia

Giardia (gee-ar-dee-ah) *lamblia* (also known as hiker's/traveller's scourge or disease) is a microscopic parasite that can infect warm-blooded animals and humans. Although *Giardia* was discovered in the 19th century, it was not until 1981 that the World Health Organization (WHO) classified *Giardia* as a pathogen. *Giardia* is protected by an outer shell called a *cyst* that allows it to survive outside the body for long periods of time. If viable cysts are ingested, *Giardia* can cause the illness known as *giardiasis*, an intestinal illness which can cause nausea, anorexia, fever, and severe diarrhea.

In the United States, *Giardia* is the most commonly identified pathogen in waterborne disease outbreaks. Contamination of a water supply by *Giardia* can occur in two ways: (1) by the activity of animals in the watershed area of the water supply, or (2) by the introduction of sewage into the water supply. Wild and domestic animals have been shown to be major contributors in contaminating water supplies. Studies have also shown that, unlike many other pathogens, *Giardia* is not host specific. This means that *Giardia* cysts excreted by animals can infect and cause illness in humans. Additionally, several major outbreaks of waterborne *Giardia* have been found to be caused by sewage contaminated water supplies.

Waterborne *Giardia*, however, can be effectively controlled by treating the water supply. Chlorine and ozone are examples of two disinfectants known to effectively kill *Giardia* cysts. Filtration of the water is also effective by trapping and removing the parasite from the water supply. The combination of disinfection and filtration is the most effective water treatment process available today.

In drinking water, *Giardia* is regulated under the Surface Water Treatment Rule. Although the SWTR does not establish a Maximum Contaminant Level (MCL) for *Giardia*, it does specify treatment requirements to achieve at least 99.9% (log 3) removal and/or inactivation of *Giardia*. This regulation requires that all drinking water systems, using surface water or groundwater under the influence of surface water, must disinfect and filter the water. The Enhanced Surface Water Treatment Rule (ESWTR), which includes *Cryptosporidium* and further regulates *Giardia*, was established in December 1996.

Giardiasis[1]

During the past fifteen years giardiasis has been recognized as one of the most frequently occurring waterborne diseases in the United States. *Giardia lamblia* have been discovered in the U.S. in places as far apart as Estes Park, Colorado (near the Continental Divide); Missoula, Montana; Wilkes-Barre, Scranton, and Hazleton, Pennsylvania; and Pittsfield and Lawrence, Massachusetts, just to name a few.

Giardiasis is characterized by intestinal symptoms that usually last one week or more and may be accompanied by one or more of the following: diarrhea, abdominal cramps, bloating, flatulence, fatigue, and weight loss. Although vomiting and fever are commonly listed as relatively frequent symptoms, they have been uncommonly reported by people involved in waterborne outbreaks in the U.S.

While most *Giardia* infections persist only for one or two months, some people undergo a more chronic phase, which can

---

[1]Much of the information contained in this section is from the U.S. Centers for Disease Control, *Giardiasis*, by Juranek, D. D., 1995.

follow the acute phase or may become manifest without an antecedent acute illness. The chronic phase is characterized by loose stools, and increased abdominal gassiness with cramping, flatulence, and burping. Fever is not common, but malaise, fatigue, and depression may occur (Weller, 1985). For a small number of people, the persistence of infection is associated with the development of marked malabsorption and weight loss (Weller, 1985). Similarly, lactose (milk) intolerance can be a problem for some people. This can develop coincidentally with the infection or be aggravated by it, causing an increase in intestinal symptoms after ingestion of milk products.

Some people may have several of these symptoms without evidence of diarrhea, or have only sporadic episodes of diarrhea every three or four days. Still others may not have any symptoms at all. Therefore, the problem may not be whether you are infected with the parasite or not, but how harmoniously you both can live together, or how to get rid of the parasite (either spontaneously or by treatment) when the harmony does not exist or is lost.

Giardiasis occurs worldwide. In the U.S., *Giardia* is the parasite most commonly identified in stool specimens submitted to state laboratories for parasitologic examination. During a three year period, approximately 4% of 1 million stool specimens submitted to state laboratories were positive for *Giardia* (CDC, 1979). Other surveys have demonstrated *Giardia* prevalence rates ranging from 1 to 20% depending on the location and ages of persons studied. Giardiasis ranks among the top 20 infectious diseases that cause the greatest morbidity in Africa, Asia, and Latin America (Walsh & Warren, 1979); it has been estimated that about 2 million infections occur per year in these regions (Walsh, 1981).

People who are at highest risk for acquiring *Giardia* infection in the U.S. may be placed into five major categories:

1. People in cities whose drinking water originates from streams or rivers and whose water treatment process does not include filtration, or the filtration is ineffective because of malfunctioning equipment
2. Hikers/campers/outdoorspeople

3. International travelers
4. Children who attend day-care centers, day-care center staff, and parents and siblings of children infected in day-care centers
5. Homosexual men

People in categories 1, 2, and 3 have in common the same general source of infections, i.e., they acquire *Giardia* from fecally contaminated drinking water. The city resident usually becomes infected because the municipal water treatment process does not include the filter necessary to physically remove the parasite from the water. The number of people in the U.S. at risk (i.e., the number who receive municipal drinking water from unfiltered surface water) is estimated to be 20 million. International travelers may also acquire the parasite from improperly treated municipal waters in cities or villages in other parts of the world, particularly in developing countries. In Eurasia, only travelers to Leningrad appear to be at increased risk. In prospective studies, 88% of U.S. and 35% of Finnish travelers to Leningrad, who had negative stool tests for *Giardia* on departure to the Soviet Union, developed symptoms of giardiasis and had positive tests for *Giardia* after they returned home (Brodsky et al., 1974; Jokipii et al., 1974). With the exception of visitors to Leningrad, however, *Giardia* has not been implicated as a major cause of traveler's diarrhea. The parasite has been detected in fewer than 2% of travelers who develop diarrhea. Hikers and campers risk infection every time they drink untreated raw water from a stream or river.

Persons in categories 4 and 5 become exposed through more direct contact with feces of an infected person, e.g., exposure to soiled diapers of an infected child (day-care center–associated cases), or through direct or indirect anal-oral sexual practices in the case of homosexual men.

Although community waterborne outbreaks of giardiasis have received the greatest publicity in the U.S. during the past decade, about half of the *Giardia* cases discussed with staff of the Centers for Disease Control (CDC) over a three year period had a day-care center exposure as the most likely source of infection. Numerous outbreaks of *Giardia* in day-care centers have been reported in recent years. Infection rates for children in day-care center outbreaks range from 21 to 44% in the U.S. and from 8 to 27% in Canada (Black et al., 1981; Pickering et al., 1984; Sealy and Schuman, 1983; Pickering et al., 1981; Keystone et al., 1984; Keystone et al., 1978). The highest infection rates are usually observed in children who wear diapers (one to three years of age). In a study of 18 randomly selected day-care centers in Atlanta (CDC unpublished data), 10% of diapered children were found to be infected. Transmission from this age group to older children, day-care staff, and household contacts is also common. About 20% of parents caring for an infected child become infected.

To understand the finer aspects of *Giardia* transmission and strategies for control, the drinking water practitioner must become familiar with several aspects of the parasite's biology. Two forms of the parasite exist: a *trophozoite* and a *cyst*, both of which are much larger than bacteria (see Figure 6.1). Trophozoites live in the upper small intestine where they attach to the intestinal wall by means of a disc-shaped suction pad on their ventral surface. Trophozoites actively feed and reproduce at this location. At some time during the trophozoite's life, it releases its hold on the bowel wall and floats in the fecal stream through the intestine. As it makes this journey, it undergoes a morphologic transformation into an egg-like structure called a cyst. The cyst, which is about 6 to 9 micrometers in diameter × 8 to 12 micrometers (1/100 millimeter) in length, has a thick exterior wall that protects the parasite from the harsh elements that it will encounter outside the body. This cyst form of the parasite is infectious for other people or animals. Most people become infected either directly by hand-to-mouth transfer of cysts from the feces of an infected individual, or indirectly by drinking feces-contaminated water. Less common modes of transmission include ingestion of fecally contaminated food, and hand-to-mouth transfer of cysts after touching a fecally contaminated surface. After the cyst is swallowed, the trophozoite is liberated through the action of stomach acid and digestive enzymes and becomes established in the small intestine.

Although infection after the ingestion of only one *Giardia*

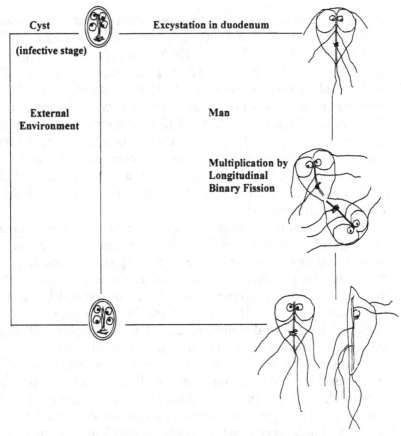

FIGURE 6.1 *Life cycle of* Giardia lamblia.

cyst is theoretically possible, the minimum number of cysts shown to infect a human under experimental conditions is 10 (Rentorff, 1954). Trophozoites divide by binary fission about every 12 hours. What this means in practical terms is that if a person swallowed only a single cyst, reproduction at this rate would result in more than 1 million parasites 10 days later, and 1 billion parasites by day 15.

The exact mechanism by which *Giardia* causes illness is not yet well understood, but is not necessarily related to the number of organisms present. Nearly all of the symptoms, however, are related to dysfunction of the gastrointestinal tract. The parasite rarely invades other parts of the body, such as the gall bladder or pancreatic ducts. Intestinal infection does not result in permanent damage.

Scientific knowledge about what is required to kill or remove *Giardia* cysts from a contaminated water supply has increased considerably. For example, it is known that cysts can survive in cold water (4°C) for at least two months and that they are killed instantaneously by boiling water (100°C) (Frost et al., 1984; Bingham et al., 1979). It is not known how long the cysts will remain viable at other water temperatures (e.g., at 0°C or in a canteen at 15–20°C), nor is it known how long the parasite will survive on various environmental surfaces, e.g., under a pine tree, in the sun, on a diaper-changing table, or in carpets in a day-care center.

The effect of chemical disinfection, such as chlorine, on the viability of *Giardia* cysts is an even more complex issue. It is clear from the number of waterborne outbreaks of *Giardia* that have occurred in communities where chlorine was employed as a disinfectant that the amount of chlorine used routinely for municipal water treatment is not effective against *Giardia* cysts. These observations have been confirmed in the laboratory under experimental conditions (Jarroll et al., 1979; Jarroll et al., 1980; Jarroll et al., 1981). This does not mean, however, that chlorine does not work at all. It does work under certain favorable conditions. Without getting too technical, one can gain some appreciation of the problem by understanding a few of the variables that influence the efficacy of chlorine as a disinfectant.

1. Water pH: at pH values above 7.5, the disinfectant capability of chlorine is greatly reduced.
2. Water temperature: the warmer the water, the higher the efficacy. Thus, chlorine does not work in ice-cold water from mountain streams.
3. Organic content of the water: mud, decayed vegetation, or other suspended organic debris in water chemically combines with chlorine making it unavailable as a disinfectant.
4. Chlorine contact time: the longer *Giardia* cysts are exposed to chlorine, the more likely it is that the chemical will kill them.

5. Chlorine concentration: the higher the chlorine concentration, the more likely chlorine will kill *Giardia* cysts. Most water treatment facilities try to add enough chlorine to give a free (unbound) chlorine residual at the customer tap of 0.5 mg per liter of water.

The long-term solution to the problem of municipal water-borne outbreaks of giardiasis involves improvements in, and more widespread use of, filters in the municipal water treatment process. The sand filters most commonly used in municipal water treatment today cost millions of dollars to install. This makes them unattractive for many small communities. Moreover, the pore sizes in these filters are not sufficiently small to remove a *Giardia* cyst (6 to 9 micrometers × 8 to 12 micrometers). For the sand filter to remove *Giardia* cysts from the water effectively, the water must receive some additional treatment before it reaches the filter. In addition, the flow of water through the filter bed must be carefully regulated.

Cryptosporidium

In 1907, when Ernest E. Tyzzer recognized, described, and published an account of a parasite he frequently found in the gastric glands of laboratory mice, his new discovery, along with himself, was hidden in virtual anonymity—just another scientist quietly going about his normal, tedious, out-of-the-limelight research, buried (along with his work) in obscurity. Initially his studies focused on describing the asexual and sexual stages and spores (oocysts), each with a specialized attachment organelle, and noted that spores were excreted in the feces (Tyzzer, 1907). Tyzzer identified the parasite as a sporozoan, but of uncertain taxonomic status; he named it *Cryptosporidium muris*. Later, in 1910, after more detailed study, he proposed *Cryptosporidium* as a new genus and *C. muris* as the type species. Amazingly, except for developmental stages, Tyzzer's original description of the life cycle (see Figure 6.2) was later confirmed by electron microscopy. Later, Tyzzer (1912) described a new species, *Cryptosporidium parvum*.

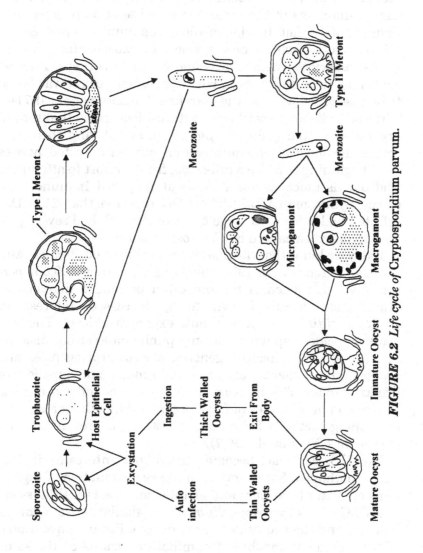

FIGURE 6.2 Life cycle of Cryptosporidium parvum.

For almost 50 years, Tyzzer's discovery of the genus *Cryptosporidium* (because it appeared to be of no medical or economic importance) remained relatively obscure. However, slight rumblings of the importance of the obscure scientist's findings were felt in the medical community when Slavin (1955) wrote about a new species, *Cryptosporidium meleagridis*, associated with illness and death in turkeys. Interest continued to be slight, even when *Cryptosporidium* was found to be associated with bovine diarrhea (Paniciera et al., 1971).

It was 1982 before worldwide interest focused on the study of organisms in the genus *Cryptosporidium*. During this timeframe the medical community and other interested parties were beginning to make a full-scale, frantic effort to attempt to find out as much as possible about Acquired Immune Deficiency Syndrome (AIDS). The CDC reported that 21 AIDS-infected males from six large cities in the U.S. had severe protracted diarrhea caused by *Cryptosporidium*.

Today we know that the massive waterborne outbreak in Milwaukee, Wisconsin, in 1993, involving more than 400,000 persons who developed acute and often prolonged diarrhea or other gastrointestinal symptoms, increased interest in *Cryptosporidium* at a new almost exponential level. The Milwaukee incident spurred not only public interest but also the interest of public health agencies, agricultural agencies and groups, environmental agencies and groups, and suppliers of drinking water. This increase in interest level and concern has spurred on new studies of *Cryptosporidium*, with emphasis on developing methods for recovery, detection, prevention, and treatment (Fayer et al., 1997).

The U.S. EPA has become particularly interested in this "new" pathogen. For example, in the reexamination of regulations on water treatment and disinfection, the U.S. EPA issued a MCLG for *Cryptosporidium*. The similarity to *Giardia lamblia*, and the necessity to provide an efficient conventional water treatment capable of eliminating viruses at the same time, forced the U.S. EPA to regulate surface water supplies in particular. The Enhanced Surface Water Treatment Rule (ESWTR) was proposed including regulations from watershed protection to specialized operation of treatment plants (certification of operators and state overview) and effective chlorination. Protection against *Cryptosporidium* includes control of waterborne pathogens, such as *Giardia* and viruses (De Zuane, 1997).

Cryptosporidium (crip-toe-spor-ID-ee-um) is one of several single-celled protozoan genera in the phylum Apircomplexa (all referred to as coccidia). The parasite lives in the intestines of animals and people. That is, *Cryptosporidium* develops in the gastrointestinal tract of vertebrates through all of their life cycle. This microscopic pathogen causes a disease called cryptosporidiosis (crip-toe-spor-id-ee-O-sis).

The dormant (inactive) form of *Cryptosporidium*, called an oocyst (O-o-sist; shown in Figure 6.2), is excreted in the feces (stool) of infected humans and animals. The tough-walled oocysts survive under a wide range of environmental conditions.

Several species of *Cryptosporidium* were incorrectly named after the host in which they were found; subsequent studies have invalidated many species. At the present time there are eight valid named species of *Cryptosporidium* (see Table 6.1).

Upton (1997) reports that *C. muris* infects the gastric glands of laboratory rodents and several other mammalian species, but (even though several texts state otherwise) is not known to infect humans. However, *C. parvum* infects the small intestine of an unusually wide range of mammals, including humans, and is the zoonotic species responsible for human cryptosporidiosis.

Table 6.1. Valid named species of Cryptosporidium.

Species	Host
C. baileyi	Chicken
C. felis	Domestic cat
C. meleagridis	Turkey
C. muris	House mouse
C. nasorum	Fish
C. parvum	House mouse
C. serpentis	Corn snake
	Rat snake
	Madagascar boa
C. wrairi	Guinea pig

Source: Adapted from Fayer et al., The General Biology of Cryptosporidium. In Cryptosporidium and Cryptosporidiosis, Fayer, R. (ed.), Boca Raton, Florida: CRC Press, 1997

Upton goes on to explain that *C. parvum* is predominately a parasite of neonate (newborn) animals. He points out that even though exceptions occur, older animals generally develop poor infections, even when unexposed previously to the parasite. But this is not the case with humans, however; they are the one host that can be infected at any time in their lives, and only previous exposure to the parasite results in either full or partial immunity to challenge infections.

Oocysts are present in most surface bodies of water (e.g., lakes and rivers) across the U.S., many of which supply public drinking water. Oocysts are more prevalent in surface waters when heavy rains increase runoff of wild and domestic animal wastes from the land, or when sewage treatment plants are overloaded or break down.

Only laboratories with specialized capabilities can detect the presence of *Cryptosporidium* oocysts in water. Unfortunately, present sampling and detection methods are unreliable. It is difficult to recover oocysts trapped on the material used to filter water samples. Looking at a sample under a microscope, it is easy to determine whether the oocyst is alive, or whether it is the species *C. parvum* that can infect humans.

The number of oocysts detected in raw (untreated) water varies with location, sampling time and laboratory methods. Water treatment plants remove most, but not always all, oocysts sufficient to cause cryptosporidiosis, but the low numbers of oocysts sometimes present in drinking water are not considered cause for alarm in the general public.

To protect water supplies from *Cryptosporidium*, multiple barriers are needed. Why? *Cryptosporidium* oocysts have tough walls that can withstand many environmental stresses, and are resistant to the chemical disinfectants, such as chlorine, that are traditionally used in municipal drinking water systems and swimming pools.

Physical removal of particles, including oocysts, from water by filtration is an important step in the municipal water treatment process. Typically, water pumped from rivers or lakes into a treatment plant is mixed with coagulants that help settle out particles suspended in the water. If sand filtration is used, even more particles are removed. Finally, the clarified water is disinfected and piped to customers. Filtration is the only conventional method now in use in the United States for controlling *Cryptosporidium*.

Cryptosporidiosis

Since the Milwaukee outbreak, concern about the safety of drinking water in the U.S. has increased, and much of the attention has been focused on determining and reducing the risk of cryptosporidiosis from community and municipal water supplies. Cryptosporidiosis is caused by infection with or by *Cryptosporidium* oocysts.

Cryptosporidiosis is spread by putting something in the mouth that has been contaminated with the stool of an infected person or animal. In this way, people swallow the *Cryptosporidium* parasite. A person can become infected by drinking contaminated water or eating raw or undercooked food contaminated with *Cryptosporidium* oocysts; direct contact with the droppings of infected animals or stool of infected humans; or hand-to-mouth transfer of oocysts from surfaces that may have become contaminated with microscopic amounts of stool from an infected person or animal.

After exposure, the symptoms may appear two to ten days after infection by the parasite. Although some persons may not have symptoms, others have watery diarrhea, headache, abdominal cramps, nausea, vomiting, and low-grade fever. These symptoms may lead to weight loss and dehydration.

In otherwise healthy persons, these symptoms usually last one to two weeks, at which time the immune system is able to stop the infection. In persons with suppressed immune systems, such as persons who have AIDS or recently have had an organ or bone marrow transplant, the infection may continue and become life-threatening.

At the present time there is no safe and effective cure for cryptosporidiosis. People who have normal immune systems improve without taking antibiotic or antiparasitic medications. The treatment recommended for this diarrhea illness is to drink plenty of fluids and to get extra rest. Physicians may prescribe medication to slow the diarrhea during recovery.

The best way to prevent cryptosporidiosis is:

- Avoid water or food that may be contaminated.
- Wash hands after using the toilet and before handling food.
- If you work in a child-care center where you change diapers, be sure to wash your hands thoroughly with plenty of soap and warm water after every diaper change, even if you wear gloves.

Classification

Protozoa are divided into four groups based on the method of motility. The *Mastigophora* are motile by means of one or more flagella; the *Ciliophora* by means of shortened modified flagella called cilia; the *Sarcodina* by means of amoeboid movement; and the *Sporozoa,* which are nonmotile. Table 6.2 lists all four groups, but in this text only the first three, Mastigophora, ciliates, and Sarcodina are discussed in detail.

Mastigophora (Flagellates)

These protozoans are mostly unicellular, lack specific shape (have an extremely flexible plasma membrane that allows for the flowing movement of cytoplasm), and possess whiplike structures called flagella. The flagella, which move with a whiplike motion, are used for locomotion, as sense receptors, and to attract food. As indicators in the wastewater treatment process, flagellates are normally associated with poor treat-

Table 6.2. Classification of protozoans.

Group	Common Name	Means of Movement	Method of Reproduction
Mastigophora	Flagellates	Flagella	Asexual
Ciliophora	Ciliates	Cilia	Asexual by transverse fission; sexual by conjugation
Sarcodina	Amoebas	Pseudopodia	Asexual and sexual
Sporozoa	Sporozoans	Nonmotile	Asexual and sexual

ment and a young biomass. When they are the predominate protozoa, the plant effluent will contain large amounts of suspended solids and BOD.

These organisms are common in both fresh and marine waters. The group is subdivided into the *Phytomastigophrea*, most of which contain chlorophyll and are thus plantlike. A characteristic species of Phytomastigophrea, the *Euglena* sp., is often associated with high or increasing levels of nitrogen and phosphate in the treatment process. A second subdivision of Mastigophora is the animal-like and nonpigmented *Zoomastigophrea*.

Ciliophora (Ciliates)

The ciliates are the most advanced and structurally complex of all protozoans. Movement and foodgetting are accomplished with short hairlike structures called cilia, which are present in at least one stage of the organism's life cycle. There are three groups of ciliates: (1) free swimmers, (2) crawlers, and (3) stalked. The majority are free swimming. They are usually solitary but some are colonial and others are sessile. They are unique among protozoa in having two kinds of nuclei: a micronucleus and a macronucleus. The micronucleus is concerned with sexual reproduction. The macronucleus, on the other hand, is involved with metabolism and the production of RNA for cell growth and function.

Ciliates are covered by a *pellicle*, which may act as a thick armor or be very thin. The cilia are short and usually arranged in rows. Their structure is comparable to flagella except that cilia are shorter. Cilia may cover the surface of the animal or may be restricted to banded regions.

Free-swimming ciliates are very specialized because they have organelles that carry out particular vital processes. Probably the best known and most widely distributed of the ciliates is the *paramecium* (Figure 6.3). The paramecium is a free swimmer with cilia of uniform length. When it comes to feeding, the paramecium is the least efficient of the three types of ciliates.

Figure 6.3 shows that the paramecium has several specialized organelles. Excretion and osmoregulation are the function

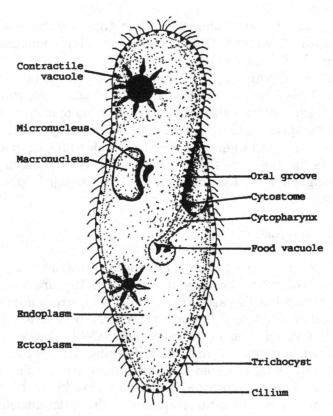

FIGURE 6.3 *Paramecium, showing cytopharynx (gullet), nuclei, contractile vacuole, and food vacuoles.*

of the *contractile vacuole*. This organelle separates a dilute solution of water and electrolytes from the cytoplasm and eventually expels this solution through the *contractile vacuole pore*. A diaphragm keeps the vacuole closed during the expanding period.

Another important organelle shown in Figure 6.3 is the *trichocyst*. When discharged, they form long threadlike filaments that pass through the pellicle and harden. The tips are sticky, which allows for attachment or acts as tools to capture and paralyze the prey prior to ingestion.

Crawler-type ciliates are also known as creeping ciliates. They creep over and slowly through floc particles. The creeping ciliates feed much more efficiently than the free swimmers and are able to survive with lowering numbers of bacteria.

Stalked-type ciliates have an enlarged disc-shaped mouth on the anterior end of their bodies. The stalk portion of their bodies usually contains a *myoneme* (a contractible filament) that resembles a corkscrew. When turbulent or low-oxygen conditions are present, they may swim around freely. This degree of freedom allows them to feed efficiently. One type of stalked ciliate has tentacles (*Actinea* sp.) and is a predator.

In wastewater treatment, two forms of ciliates are important: the free-swimming and stalked ciliates. The free-swimming ciliates use their cilia to move through the wastewater and to attract food. They are normally associated with a moderate sludge age and effluent quality. When they are the predominate organisms, they are indicated by a plant effluent that is turbid and contains a high amount of suspended solids [see Figure 6.4(a)].

The stalked ciliates attach themselves to wastewater solids and use their cilia to attract food. These ciliates are normally associated with a plant effluent that is very clear and contains low amounts of suspended solids and BOD [see Figure 6.4(b)].

Sarcodina

Members of this group have fewer organelles and are simpler in structure than the ciliates and flagellates. Sarcodina move

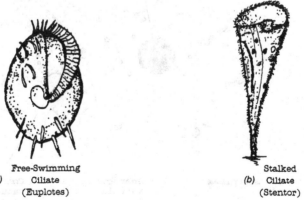

	Free-Swimming		Stalked
(a)	Ciliate	(b)	Ciliate
	(Euplotes)		(Stentor)

FIGURE 6.4 (a) Shows a free-swimming ciliate; (b) shows a stalked ciliate.

about by the formation of flowing protoplasmic projections called *pseudopodia*. The formation of pseudopodia is commonly referred to as *amoeboid movement*. The amoebas are well known for this mode of action (see Figure 6.5). The pseudopodia not only provide a means of locomotion but also serve as a means of feeding, which is accomplished when the organism puts out the pseudopodium to enclose the food. Most amoebas feed on algae, bacteria, protozoa, and rotifers.

In wastewater treatment, amoebas are associated with poor treatment or a young biomass and (normally) with treatment that produces an effluent high in suspended solids and BOD. Amoebas move through wastewater by moving the liquids stored in their cell wall.

Several different wastewater treatment processes have large numbers of protozoans. These processes include activated sludge, oxidation ponds, rotating biological collectors (RBCs), trickling filters, and wetlands. Protozoans thrive in aerobic processes for several reasons: increased levels of oxygen, high organic content of wastewater, and large populations of bacteria are all factors that contribute to growth of protozoa in wastewater treatment processes.

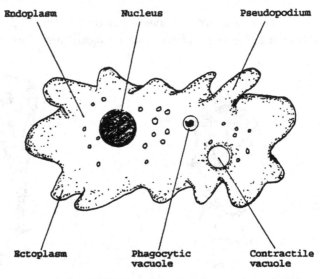

FIGURE 6.5 *Amoeba.*

As pointed out earlier, protozoans, along with acting as parameters of conditions existing in the treatment process, play a beneficial role in wastewater treatment. For example, their removal of or cropping of bacteria aids in clarification of secondary effluent. This clarification is accomplished through flocculation of suspended and colloidal materials. Protozoa also aid in the nitrification process. As biological indicators of process quality and health, they are especially indicative of the quality of secondary effluent and also of the health of activated sludge.

Activated Sludge

The activated sludge process originated in England. It is so named because it is involved in the production of an activated mass of microorganisms capable of aerobically stabilizing the organic content of the waste. Activated sludge is a biological process where contact with bacteria, protozoa, fungi, and other small organisms such as rotifers and nematodes occurs. The bacteria are the most important group of microorganisms for they are the ones responsible for the structural and functional activity of the activated sludge floc. In this process it is important to bring the bacteria and other microorganisms in contact with the organic matter in the wastewater. This is usually accomplished through rapid mixing. This mixing or agitation process, provided by large mixers, is augmented by aeration. Agitation and aeration work hand-in-hand to mix the returned sludge with effluent from primary treatment, to keep the activated sludge in suspension, and to supply oxygen to the biochemical reactions necessary for the stabilization of the wastewater.

The activated sludge process is typified by the successive development of protozoa and mature floc particles. This succession can be indicated by the presence of the type of dominant protozoa present. At the start of the activated sludge process (or recovery from an upset condition), the amoebas dominate. As the process continues, uninterrupted or without upset, small populations of bacteria begin to grow in logarithmic fashion, which, as the population increases, develop into mixed liquor. When this occurs, the flagellates dominate. When the sludge attains an age of about three days, lightly dispersed floc particles begin to form and there is an increase in bacteria. At this point, the free-swimming ciliates dominate. The process goes on with floc particles beginning to stabilize, taking on irregular shapes, and starting to show filamentous growth. At this stage, the crawling ciliates dominate. Eventually, mature floc particles develop, increase in size, and large numbers of crawling and stalked ciliates are present. When this occurs, the succession process has reached its terminal point.

The succession of protozoan and mature floc particle development just described details the occurrence of phases of development in a step-by-step progression. This is also the case when protozoan succession is based on other factors such as dissolved oxygen and food availability.

Probably the best way in which to understand protozoan succession based on dissolved oxygen and food availability is to view the wastewater treatment plant's aeration basin as a "stream within a container." Using the *saprobity system* to classify the various phases of the activated sludge process in relation to the self-purification process that takes place in a stream, one is able to see a clear relationship between the two processes based on available dissolved oxygen and food supply. To gain an understanding of this relationship and process, consider the following explanation.

The stream self-purifies and stabilizes over distance. That is, as the polluted (low dissolved oxygen and food supply) stream flows from point of pollution, it stabilizes. In the aeration basin, on the other hand, waste stabilization is based on time (sludge age) not distance. Each condition (except the last phase) described in the saprobity system is experienced in the aeration basin.

In the stream, a number of distinct zones of pollution can be identified according to their degree of pollution, content of dissolved oxygen, and the types of biotic indicators that are present. The zones used to describe the extent of these conditions are as follows:

1. *Polysaprobic zone*—point in the stream where pollution occurs and dissolved oxygen declines

2. *α-Mesoprobic zone*—point in the stream where pollution is heavy and dissolved oxygen is low
3. *β-Mesoprobic zone*—point in the stream where pollution is moderate and dissolved oxygen is increasing
4. *Oligosaprobic zone*—point in the stream where pollution is low and dissolved oxygen levels are almost normal
5. *Xenosaprobic zone*—point in the stream where pollution is nonexistent and dissolved oxygen is normal

As stated previously, except for the last zone, xenosaprobic zone, the other zones and their associated conditions of pollution and dissolved oxygen contents are similar to the environment within the activated sludge aeration basins.

Competition for Food in Activated Sludge Systems

Any change in the relative numbers of bacteria in the activated sludge process has a corresponding change to microorganism population. Decreases in bacteria increase competition between protozoa and result in succession of dominant groups of protozoans.

The degree of success or failure of protozoa to capture bacteria depends on several factors. For example, those with more advanced locomotion capability are able to capture more bacteria. Individual protozoan feeding mechanisms are also important in the competition for bacteria. At the beginning of the activated sludge process, amoebas and flagellates are the first protozoan groups to appear in large numbers. They can survive on smaller quantities of bacteria because their energy requirements are lower than other protozoan types. Since there is little bacteria present, competition for dissolved substrates is low. However, as the bacteria population increases, these protozoans are not able to compete for available food. This is when the next group of protozoans, the free-swimming protozoans, enter the scene.

The free-swimming protozoans take advantage of the large populations of bacteria because they are better equipped with food-gathering mechanisms than the amoebas and flagellates.

The free swimmers are important not only because of their insatiable appetites for bacteria but also in floc formation. Secreting polysaccharides and mucoproteins that are absorbed by bacteria—which make the bacteria "sticky" through biological agglutination (biological gluing together)—allows them to stick together and, more importantly, to stick to floc. Thus, large quantities of floc are prepared for removal from secondary effluent and are either returned to aeration basins or wasted.

The crawlers and stalked ciliates succeed the free swimmers. The free swimmers are replaced in part because of the increasing level of mature floc retards their movement. Additionally, the type of environment that is provided by the presence of mature floc is more suited to the needs of the crawlers and stalked ciliates. The crawlers and stalked ciliates also aid in floc formation by adding weight to floc particles, thus enabling removal.

From the preceding discussion it is evident that protozoa are important members of the microorganism population of the activated sludge process in wastewater treatment. They not only consume, and thus remove, bacteria from the activated sludge and secondary effluent but also help with nitrification. In addition, the protozoans act as parameters of sludge health and effluent quality. By simple examination and identification of the protozoan population in activated sludge, it is possible to determine whether or not loading is at acceptable or unacceptable levels. The presence of a particular species of protozoa can also indicate whether or not the process is operating correctly. The protozoan varieties can also indicate changes taking place in the strength and composition of the wastewater.

The importance of using protozoans as parameters to indicate sludge health and effluent quality cannot be overemphasized. To gain better understanding of how protozoan indicators can be used to determine the quality level of process operation, the following parameters are provided:

- Healthy sludge is indicated when large varieties of crawlers and stalked ciliates are observed. A condition such as that just described can indicate to the wastewater specialist that the process is producing a high quality effluent with BOD ranging from 1–10 mg/L.
- Intermediate sludge is indicated by a preponderance of all

three ciliated groups. When this occurs the indication is that the effluent is of satisfactory quality with a BOD ranging from 11–30 mg/L.

• Poor sludge is indicated when the population is dominated by free swimmers and flagellates. The effluent is generally turbid and of low quality with BOD at levels greater than 30 mg/L.

The indicators of effluent quality can be used in other ways as well. A significant shift from the patterns described may indicate that the sludge age is significantly high and/or the presence of excessive nutrient levels (nitrogen or phosphate). The absence of or too few protozoans in activated sludge processes can also indicate process problems. For example, when the protozoan population is too low or absent, the F/M ratio may be too high (an overloaded condition).

Environmental Factors Affecting Protozoan Population

The population, activity, and diversity of the protozoa in activated sludge and other treatment processes are affected by environmental factors. For example, the availability of nontoxic bacteria is important. The dissolved oxygen levels are important even though the protozoans are generally aerobic. Dissolved oxygen content indicates degree of pollution in the system. Toxicants like chemical surfactants affect the plasma membrane and enzyme systems of protozoans; using surfactants could also lead to the development of bacteria that are harmful to protozoa. pH level is important. Most protozoans have an optimum upper and lower pH range, shifting pH may favor one variety of protozoa over the other. Significant amounts of rainfall can affect the protozoa; that is, significant decreases in overall population can occur through hydraulic washout.

ROTIFERS

Rotifers make up a well-defined group of the smallest, simplest multicellular microorganisms and are found in nearly all aquatic habitats. These strict aerobes range in size from about 0.1–0.8 nm. Often associated with aerobic biological processes in wastewater treatment plants, they are seen either grazing on bacteria or attached to debris by their forked tail or toe (see Figure 6.6). Rotifers promote microfloral activity and decomposition, enhance oxygen penetration in activated sludge and trickling filters, and recycle minerals in each. Most descriptions apply to the female because the male is much smaller and structurally simpler. Rotifers form into various shapes such as spherical, sac, and/or wormlike. Their forms are composed of three zones. At their anterior end, rotifers possess actively moving cilia that frequently beat in a circular motion for motility and food gathering. The main body form below the head possesses a thick cuticle that terminates at the foot end. The foot possesses adhesive (cement) glands and toes for attachment to substratum. Rotifers are unique in the sense that they possess the ability to chew their food by using a modified muscular pharynx called a mastax. Rotifers require high levels of dissolved oxygen; thus, their presence indicates water with a high level of biological purity.

Rotifers possess reproductive organs in the form of gonads. Moreover, they are separated into two different orders according to the number of gonads they possess. For example, in the

FIGURE 6.6 Philodina, *a common rotifer.*

order Monogononta, the rotifers possess one gonad; in the order Digononta, rotifers possess two gonads.

Movement by rotifers is accomplished either by the free-swimming or crawling mode. Free swimmers move by the beating action of rings of cilia on the epidermal area of the head. When each ring of cilia beats it gives the impression of a wheel with spokes. The frequency of this beating motion is quite high. Rotifers of this type move in a forward direction at a slow pace.

Rotifers that move using a crawling motion employ an interesting technique to accomplish their movement. While attached by its adhesive glands and toes to an old substratum the rotifer will extend its body. While extending, the rotifer's head will use its adhesive glands to attach to a new substratum. Then the toes are released from the old substratum. The body contracts so that the foot reaches around and attaches to the substratum close to the head. The head then releases and the body is extended to its normal posture.

Rotifers feed on algae, bacteria, protozoa, and dead organisms. The quantity of food that rotifers can remove from a treatment process is quite significant.

In wastewater treatment, rotifers play an important role in activated sludge, trickling filter, and oxidation pond systems. In activated sludge and trickling filter processes large numbers of both Digononta and Monogononta are present. In oxidation ponds the Monogononta predominate.

In oxidation ponds rotifers feed mainly on algae. By continuously cropping the algae, rotifers recycle nutrients. In trickling filters rotifers work to keep the biomass open, allowing for aeration of the biomass and reducing ponding of the filter. In activated sludge processes rotifers remove bacteria, encouraging active and efficient growth of more bacteria. In addition to cropping bacteria, rotifers assist in floc formation in activated sludge processes. By excreting waste and mucus consisting of bonding agents, the rotifers assist in providing bonding sites for the bacteria and other inert materials to develop upon.

In activated sludge processes rotifers are parameters that indicate process condition. For example, rotifers are strict aerobes and are only found when the environment contains at least several mg/L of dissolved oxygen. Moreover, the presence of rotifers in activated sludge can indicate an increase in the stabilization of organic wastes. On the other hand, when the rotifer population is low or nonexistent, it may indicate that the food (trophic) level is too low. Additionally, low rotifer counts may also indicate that detention time is too short. Rotifers are also temperature sensitive; that is, if the temperature is high, rotifers develop at a faster rate. On the other hand, if their environmental temperature is too low, rotifers have a correspondingly low development rate.

CRUSTACEANS

Because they are important members of freshwater zooplankton, microscopic crustaceans are of interest to water and wastewater specialists. These microscopic organisms are characterized by a rigid shell structure. They are multicellular animals that are strict aerobes, and as primary producers they feed on bacteria and algae. They are important as a source of food for fish. Additionally, microscopic crustaceans have been used to clarify algae-laden effluents from oxidation ponds. The *Cyclops*

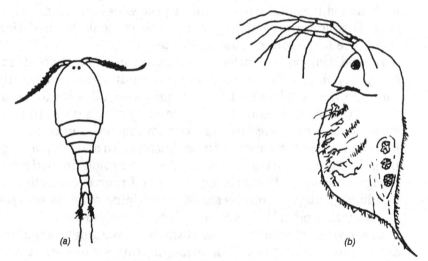

FIGURE 6.7 *(a)* Cyclops; *(b)* Daphnia.

and *Daphnia* are the two microscopic crustaceans of interest to water and wastewater operators.

Cyclops lacks the shell-like carapace (covering) of *Daphnia;* instead, they exhibit distinct body segmentation. *Cyclops* have three simple eyes or a single median eye (see Figure 6.7).

Daphnia (commonly known as the water flea) have a distinct head and a body covered by a bivalvelike carapace. They have a moveable compound eye and many appendages such as antennae, mouth parts, and four to six pairs of legs (see Figure 6.7).

WORMS (NEMATODES AND FLATWORMS)

Along with inhabiting organic muds, worms also inhabit biological slimes; they have been found in activated sludge and in trickling filter slimes (McKinney, 1962). Microscopic in size, they range in length from 0.5–3 mm and in diameter from 0.01–0.05 mm. Most species have a similar appearance. They have a body that is covered by cuticle, are cylindrical, nonsegmented, and taper at both ends.

These organisms continuously enter wastewater treatment systems, primarily through attachment to soils that reach treatment plants through inflow and infiltration. They are present in large, often highly variable numbers, but as strict aerobes they are found only in aerobic treatment processes where they metabolize solid organic matter.

Once nematodes are firmly established in the treatment process, they can promote microfloral activity and decomposition. They crop bacteria in both the activated sludge and trickling filter systems. Their activities in these systems enhance oxygen penetration by tunneling through floc particles and biofilm. In activated sludge processes they are present in relatively small numbers because the liquefied environment is not a suitable habitat for crawling, which they prefer over the free-swimming mode. In trickling filters where the fine stationary substratum is suitable to permit crawling and mating, nematodes are quite abundant.

Along with preferring the trickling filter habitat, nematodes play a beneficial role in this habitat. For example, they break loose portions of the biological slime coating the filter bed. This action prevents excessive slime growth and filter clogging. They also aid in keeping slime porous and accessible to oxygen by tunneling through slime.

In the activated sludge process, the nematodes play important roles as agents of better oxygen diffusion. They accomplish this by tunneling through floc particles. They also act as parameters of operational conditions in the process, such as low dissolved oxygen levels (anoxic conditions) and the presence of toxic wastes.

Environmental conditions have an impact on the growth of nematodes. For example, in anoxic conditions their swimming and growth is impaired. The most important condition they indicate is when the wastewater strength and composition has changed. Temperature fluctuations directly affect their growth and survival; population decreases when temperatures increase.

Aquatic flatworms (improperly named because they are not all flat) feed primarily on algae. Because of their aversion to light, they are found in the lower depths of pools. Two varieties of flatworms are seen in wastewater treatment processes: *microtubellarians* are more round than flat and average about 0.5–5 mm in size, and *macrotubellarians* (planarians) are more flat than round and average about 5–20 mm in body size. Flatworms are very hardy and can survive in wide variations in humidity and temperature. As inhabitants of sewage sludge, they play an important part in sludge stabilization and as bioindicators or parameters of process problems. For example, their inactivity or sluggishness might indicate a low dissolved oxygen level or the presence of toxic wastes.

Surface waters that are grossly polluted with organic matter (especially domestic sewage) have a fauna that is capable of thriving in very low concentrations of oxygen. A few species of tubificid worms dominate this environment. Pennak (1989) reports that the bottoms of severely polluted streams can be literally covered with a "writhing" mass of these tubificids.

The *Tubifex* (commonly known as sludge worms), are small, slender, reddish worms that normally range in length from 25

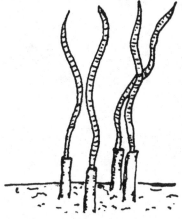

FIGURE 6.8 Tubifex.

to about 50 mm. They are burrowers; their posterior end protrudes to obtain nutrients (see Figure 6.8). When found in streams, *Tubifex* are indicators of pollution.

SUMMARY OF KEY TERMS

- *Protozoa:* protists that are defined as usually motile eucaryotic unicellular microorganisms
- *Protozoan locomotory organelles:* some have no means of locomotion, while others have pseudopodia, flagella, or cilia

CHAPTER 7

Enzymes

IN wastewater treatment, biological processes are utilized to degrade organic matter. In order to degrade organic matter, the environments in these processes must accommodate the appropriate types of microorganisms that are capable of performing the organic degradation function. In order for this to occur, the process environment must be controlled. In controlling the environment for appropriate biological growth, with the ultimate goal of decomposition of organic substances, the wastewater specialist should possess a basic understanding of enzymes and enzymatic reactions.

Enzymes, present within the microorganisms and the surrounding waste stream, are the essential biological catalysts that enable microorganisms to break down organics. A *catalyst* is a substance that modifies and increases the rate of chemical reaction without being consumed in the process.

For microorganisms to produce the enzymes needed to break down the organics they must first be acclimated to their environment. This is the case because different organics require specific enzymes to break them down. In this breaking-down process, the enzyme works to speed up the rate of hydrolysis of complex organic compounds and the rate of oxidation of simple compounds by decreasing the activation energy required (McKinney, 1962).

The living cell is the site of tremendous biochemical activity called *metabolism* (to be discussed in detail later). Metabolism is the process of chemical and physical change that goes on continually in the living organism. Conversion of food to usable en-

95

ergy, buildup of new tissue, replacement of old tissue, disposal of waste products, and reproduction are all activities that are characteristic of life.

For illustrative purposes, consider the activities that characterize life as being related to the activities that take place on the assembly line in a factory. Breslow (1990) saw this correlation when he called enzymes the "machines of life." An enzyme can repeat a particular process several times a second or even faster (like a machine) and produce consistent results. Taking this concept another step for purposes of clarity, consider the enzyme as a small machine (in this case): Raw material goes in, finished product comes out. The point is that just like a machine, an enzyme is specialized. Finally, when one considers a living cell as a small factory containing thousands of different types of specialized machines (enzymes), the function of enzymes becomes clear.

The phenomenon of *catalysis* makes possible the biochemical reactions necessary for all life processes. Catalysis is defined as the modification of the rate of a chemical reaction by a catalyst. The catalysts of biochemical reactions are enzymes; they are responsible for bringing about all of the chemical reactions in living organisms. Without enzymes, these reactions take place at a rate far too slow to keep pace with metabolism.

NATURE OF ENZYMES

What exactly are enzymes? Enzymes are essentially proteins formed by the polymerization of some or all the amino acids; twenty amino acids are found in proteins. Enzymes are high molecular weight compounds (ranging from 10,000–2,000,000) made up of chains of amino acids linked together by peptide bonds. In the overall linking process, a water molecule is removed between the carboxyl group of one amino acid and the amino group of the next one. Several steps are involved in the synthesis of proteins, including enzymes, so that chemical energy can be supplied from other molecules.

Most enzymes are pure proteins. However, other enzymes require the participation of small nonprotein groups, which may be organic or inorganic, before their catalytic activity can be ex-

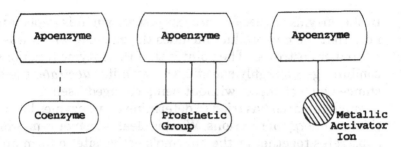

FIGURE 7.1 *Holoenzymes showing apoenzymes and various types of cofactor. [Adapted from Witkowski and Power (1975), p. 7.]*

erted. These nonprotein groups are called *cofactors* (the activator). In some cases these cofactors are nonprotein metallic ion activators (e.g., ions of iron) that form a functional part of the enzyme. When the cofactor and the protein part (the *apoenzyme*) of the enzyme are present, the entire active complex is called the *holoenzyme*. This relationship can be easily seen in the following:

$$Apoenzyme + Cofactor = Holoenzyme$$

The structural nomenclature of enzymes is affected by the way in which the cofactor is attached to the apoenzyme (see Figure 7.1). For example, if the cofactor is firmly attached to the apoenzyme, it is called a *prosthetic group*. On the other hand, when the cofactor is loosely attached to the cofactor, it is called a *coenzyme* (Prescott et al., 1993).

ACTION OF ENZYMES

It is important to keep in mind that enzymes increase the speed of reactions, without themselves undergoing any permanent chemical change (i.e., they do not alter their equilibrium constants). They are neither used up in the reaction nor do they appear as products of the reaction. This basic enzymatic reaction process can be seen in the following:

$$Substrate + Enzyme \text{ (catalyzing the reaction)} \rightarrow$$

$$Product + Enzyme$$

In the enzymatic reaction process just shown, it is important to note that the end product includes the enzyme, which was not altered or destroyed. The point is that the enzyme functions by combining in a highly specific way with its *substrate*, the substance being changed, without being changed itself.

Much research has tried to determine how enzymes lower activation energy of reactions. What is clear is that enzymes bring substrates together at the enzyme's active site to form an enzyme-substrate complex (see Figure 7.2).

In the enzyme-substrate complex the substrate is attached by weak bonds to several points in the active site of the enzyme. This bringing together of enzyme and substrate allows for their

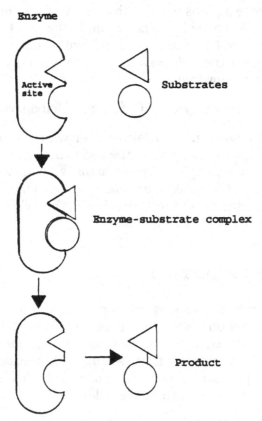

FIGURE 7.2 *Enzyme function showing the interaction of the substrate and enzyme with the resulting product. [Adapted from Prescott et al. (1993), p. 141.]*

concentration, which lowers the activation energy required to complete the reaction. It is interesting to note that most of these reactions take place at relatively low temperatures ranging from 0–36°C.

EFFICIENCY, SPECIFICITY, AND CLASSIFICATION OF ENZYMES

Enzymes are extremely efficient. Only minute quantities of an enzyme are required to accomplish at low temperatures what normally would require, by ordinary chemical means, high temperatures and powerful reagents. For example, 1 ounce of pepsin can digest almost 2 tons of egg whites in a few hours, whereas without the enzyme 15 tons of strong acid would require 36 hours at high temperature to digest the same (Witkowski & Power, 1975).

Along with being efficient and extremely reactive, enzymes are characterized by a high degree of *specificity*. That is, just as a certain key will not fit or unlock each and every lock, enzymes also require an exact molecular fit between the enzyme and the substrate.

By 1956 the number of known enzymes was rapidly increasing. In 1961 the International Union of Biochemistry published an enzyme classification scheme that is universally used today. With the exception of the originally studied enzymes such as rennin, pepsin, and trypsin, most enzyme names end in *-ase*. Standards of enzyme nomenclature, initiated by the International Union of Biochemistry, recommended that enzymes be named in terms of both the substrate acted upon and the type of reaction catalyzed.

EFFECT OF ENVIRONMENT ON ENZYME ACTIVITY

Several factors affect the rate at which enzymatic reactions proceed. These factors include substrate concentration, enzyme concentration, pH, temperature, and the presence of activators or inhibitors.

Substrate Concentration

At low substrate concentrations, an enzyme makes product slowly. However, if the amount of enzyme is kept constant and the substrate concentration is gradually increased, the reaction velocity will increase until it reaches a maximum (usually expressed in terms of the rate of product formation). After this point, increases in substrate concentration will not increase the velocity because the available enzyme molecules are binding substrate and converting it to product as rapidly as possible—the enzyme has reached the saturation point and is operating at maximal velocity.

In order to provide an in-depth understanding of this enzyme saturation process, it would be necessary to discuss in detail saturation kinetics. The study of saturation kinetics (Michaelis-Menten kinetics) is beyond the scope of this text. However, a fundamental appreciation of what is occurring during the enzyme-saturated-with-substrate phenomenon can be gained by studying the graphical representation in Figure 7.3.

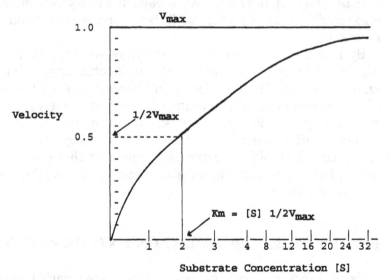

FIGURE 7.3 *The effect of substrate concentration; that is, it shows the dependence of velocity on substrate concentration for a simple on-substrate enzyme-catalyzed reaction. This substrate curve fits the Michaelis equation, which relates reaction velocity (v) to the substrate concentration (S).*

Figure 7.3 shows that the maximum velocity (V_{max}) is the rate of product formation when the enzyme is saturated with substrate and making product as fast as possible. The Michaelis constant (K_m) is the substrate concentration required for the enzyme to operate at half its maximal velocity. Theoretically, when the maximum velocity has been reached, all the available enzyme has been converted to enzyme-substrate complex. This point on the graph is designated V_{max}. By using this maximum velocity and Equation (7.1), Michaelis developed a set of mathematical expressions to calculate enzyme activity in terms of reaction speed from measurable data.

Equation (7.1) describes the simplest of several kinetic models that are consistent with saturation behavior.

$$E + S \underset{k_{-1}}{\overset{k_{+1}}{\rightleftharpoons}} ES \underset{k_{-2}}{\overset{k_{+2}}{\rightleftharpoons}} P + E \tag{7.1}$$

The Michaelis constant K_m is defined as the substrate concentration at one-half the maximum velocity. Using this constant and the fact that K_m can also be defined as

$$K_m = \frac{k_{+1} + k_{+2}}{k_{-1}} = [S]V_{max} \tag{7.2}$$

with k_{+1}, k_{-1}, and k_{+2} being the rate constants from Equation (7.1). Michaelis developed Equation (7.3) for the reaction velocity in terms of this constant and the substrate concentration.

$$V_t = \frac{V_{max}[S]}{K_m + [S]} \tag{7.3}$$

where

V_t = the velocity at any time t

$[S]$ = the substrate concentration at this time

V_{max} = the highest velocity under this set of conditions (temperature, pH, etc.)

K_m = the Michaelis constant for the enzyme being studied

Michaelis constants have been determined for many enzymes. The size of K_m indicates certain characteristics about a particular enzyme. For example, the higher the K_m value, the higher the substrate concentration at which an enzyme catalyzes its reaction. A small K_m value indicates that the enzyme requires only a small amount of substrate to become saturated. Thus, maximum velocity is reached at low substrate concentrations.

Enzyme Concentration

In order to study the effect of increasing the enzyme concentration upon the reaction rate, the reaction must be independent of the substrate concentration. Any change in the amount of product formed over a specific time frame will depend on the level of enzyme present.

pH

Enzymes also change activity with alterations in pH. The most favorable pH value (the point where the enzyme is most active) is known as the optimum pH. When the pH is much higher or lower than the enzyme's optimum value, activity slows and the enzyme is damaged.

Temperature

Enzymes also have temperature optima for maximum activity. Like most chemical reactions, the rate of an enzyme-catalyzed reaction increases with temperature. However, if the temperature rises too much above the optimum, an enzyme's structure will be denatured (disrupted) and its activity lost. The temperature optima of a microorganism's enzymes often reflect the temperature of its habitat. This is demonstrated by bacteria that grow best at high temperatures; they often have enzymes with high temperature optima.

Inhibitors

Enzyme inhibitors are substances that slow down, or in some

cases stop, catalysis. An inhibitor competes with the substrate at an enzyme's catalytic site and prevents the enzyme from forming product.

SUMMARY

In the wastewater treatment process, the goal is to have bacteria remove the material from the water by chemically transforming it into something that is separable and disposable. Enzymes play an important role in the transformation or degradation process that is accomplished when enzymes are secreted by microorganisms and decompose the organic substances. By providing sufficient substrate, mixing, and aeration in a biological system, plant operators can make the mass transfer resistances as small as possible so that the biological reaction rate controls the rate of waste removal (Sundstrom & Klei, 1979). Operator vigilance and corrective action are required in this process because of changing environmental conditions that have a direct impact upon microbial activities.

SUMMARY OF KEY TERMS

- *Enzyme:* an organic protein catalyst that causes changes in other substances without undergoing change itself
- *Catalyst:* speeds up reaction or causes a reaction to occur without changing itself
- *Coenzyme:* simpler portion of an enzyme, which is necessary for the enzyme's activation and reaction with a substrate
- *Cofactor:* a substance, such as an inorganic ion or coenzyme that activates an enzyme
- *Apoenzyme:* a protein that requires a coenzyme to function as an enzyme

CHAPTER 8

Metabolic Transformations

THE assembly-line activities that occur in microorganisms during the processing of raw materials into finished products are called metabolic transformation. Whether it be for the bacteriological control process in water treatment or for the control of microbial growth in wastewater treatment, an understanding of a microorganism's metabolic transformation is essential for water and wastewater specialists.

GENERAL METABOLISM

Metabolism is derived from the Greek word *metabole,* which means "to change." Change is what metabolism is all about. Several descriptions are used to further characterize a living organism's metabolic process(es) (its metabolism). For instance, Olomucki (1993) stated, at least in an organizational sense, that an organism's metabolism with its associated processes is its faculty to self-organize. Metabolism can also be defined as the total chemical and physical processes by which an organism maintains its functional and nutritional activities. In scientific terms, metabolism is generally referred to as the entire set of chemical reactions by which a cell produces and forms the various molecules it needs to maintain itself. In simplistic terms, metabolism can be characterized as the flow of energy through the organism.

Any definition or explanation of an organism's metabolism must also explain the metabolic processes involved. These processes are well known and well documented. For instance, the two general categories of metabolism are *catabolism* and *anabolism*. In catabolic reactions, complex compounds are broken down with a release of energy. These reactions are linked to anabolic reactions that result in the formation of important molecules. As a result of chemicals and associated reactions, biological cells are dynamic structures that are continually undergoing change.

During metabolism, the cell takes in nutrients (to be discussed later), converts them into cell components, and excretes waste into the external environment (see Figure 8.1). Microbial cells are made up of chemical substances, and when the cell grows, these chemical constituents increase in amount. The chemical substances that cells need come from the environment, that is, from outside the cell. Once inside, the cell transforms these substances into its (the cell's) basic constituents.

Metabolic reactions require energy for the uptake of various nutrients and for locomotion in motile species. Microorganisms are placed into metabolic classes based on the source of energy they use. In describing these classes the term *troph* is used (from the Greek meaning "to feed"). Thus, microorganisms that use inorganic materials as energy sources are called *litho-*

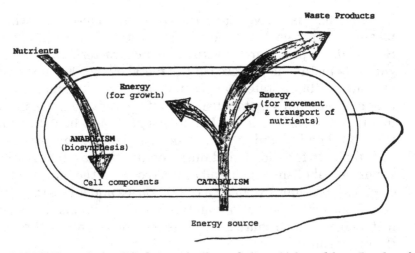

FIGURE 8.1 *A simplified view of cell metabolism. [Adapted from Brock and Madigan (1991), p. 93.]*

trophs (litho is from the Greek for rock). Microorganisms that use organic chemicals as energy sources are called *heterotrophs* (or feeding from sources other than oneself). Microorganisms that use light as an energy source are called *phototrophs* (photo is from the Greek for light). Most bacteria obtain energy from chemicals taken from the environment and are called *chemotrophs*.

Although an in-depth discussion of the metabolic processes of all microorganisms is beyond the scope of this text, water and wastewater specialists need to be well grounded in the fundamental concepts that will be covered in brief fashion in the following discussion. In particular, information will be provided so that cell metabolism and the basics of biochemistry of microbial growth can be understood. These concepts are important. Moreover, a basic understanding of microorganism metabolic processes will aid water and wastewater specialists in developing useful laboratory procedures for culturing microorganisms (to be covered in Part 2) and in developing procedures for preventing the growth of unwanted microbes.

Changes in energy accompany the chemical reactions that occur in cells. A chemical reaction can occur with the release of free energy, in which case it is called *exergonic*, or with the consumption of free energy, in which case it is called *endergonic*. The free energy of these reactions can be expressed quantitatively.

Before a chemical reaction can take place, the reactants in a chemical reaction must be activated. This activation requires energy. The amount of activation energy can be decreased by the use of a catalyst. The catalysts of living cells are enzymes, and as stated in the preceding chapter, enzymes are proteins, which are highly specific in the reaction that they catalyze.

The utilization of chemical energy in living organisms involves *oxidation-reduction* reactions—which involves the transfer of electrons from one reactant to another. *Oxidation* is defined as the removal of an electron or electrons from a substance. A *reduction* is defined as an addition of an electron (or electrons) to a substance. In the oxidation-reduction reaction, a transfer of electrons from one reactant to another takes place. The energy source, which is the electron donor, moves up one or more electrons, which are transferred to an electron acceptor. In this process, the electron donor is oxidized and the electron acceptor is reduced. One of the most common electron acceptors of living organisms is molecular oxygen. The dependency of a compound to accept or release electrons is expressed quantitatively by its reduction potential.

The transfer of electrons from donor to acceptor in a cell involves one or more intermediates, referred to as electron carriers. Some electron carriers are freely diffusible, transferring electrons from one place to another in the cell; others are firmly attached to enzymes in the cell membrane.

Two of the most common electron carriers are the coenzymes NAD and NADP. NAD+ (nicotinamide-adenine dinucleotide) and NADP+ (NAD-phosphate) are freely diffusible carriers of hydrogen atoms and always transfer two hydrogen atoms to the next carrier in the chain.

In most cases, biological reactions are catalyzed by specific enzymes that can react only with a limited range of substrates. Oxidation-reduction reactions may be considered to proceed in three stages: (1) removal of electrons from the primary donor, (2) transfer of electrons through a series of electron carriers, and (3) addition of electrons to the terminal acceptor. Each step in the reaction is catalyzed by a different enzyme, each of which binds to its substrate and to its specific coenzyme. After a coenzyme has performed its chemical functions in one reaction, it can diffuse through the cytoplasm until it attaches to another enzyme that requires the coenzyme to go back to its original form; then the process can be repeated.

Neither chemicals from the environment nor sunlight can be used directly to fuel a cell's energy-requiring processes. Therefore, the cell must have ways of converting sources of energy into a usable form of energy. In the presence of sunlight and certain chemicals, cells can make specific high-energy compounds with which to satisfy their energy demands; one of these important compounds is adenosine triphosphate (ATP).

The process of making ATP involves combining adenosine diphosphate (ADP) and inorganic phosphate (Pi) as shown in the following:

$$ADP + Pi + Energy \rightarrow ATP$$

The energy required for this reaction can be obtained in three ways—*photosynthetic phosphorylation* (the changing of an organic substance into an organic phosphate), *substrate phosphorylation,* or *oxidative phosphorylation* (occurs on the membranes of mesosomes and related structures of procaryotes)—depending upon the source of energy (see Figure 8.2).

In photosynthetic phosphorylation, the required amount of energy is absorbed by chlorophyll as light. For example, photo-

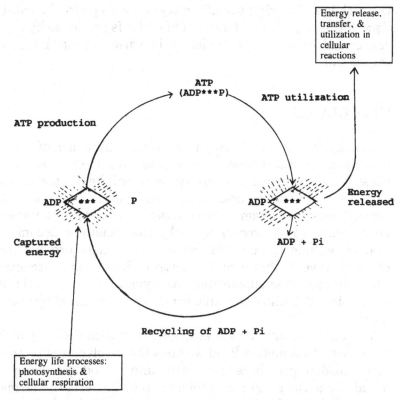

FIGURE 8.2 *The formation of ATP, a substance that fuels all living organisms. By means of phosphorylation, energy-rich bonds (***) are formed and used to combine ADP and Pi into ATP—which is used to fuel the life processes. Then the Pi and ADP are used again in a continuous cycle. [Adapted from Wistreich and Lechtman (1980), p. 273.]*

synthesis supplies the blue-green bacteria, algae, and plants with the ATP needed for synthesis (the formation of a compound from its constituents) of all materials.

The catabolic reactions by which organic compounds are converted into other organic compounds are called substrate reactions. As stated earlier, a substrate is the substance acted upon by the enzyme. During some substrate reactions, energy-rich bonds are formed, and the energy can be used to combine ADP and inorganic phosphates into ATP. These molecules of ATP are formed by substrate phosphorylation, which occurs in the cytoplasm of cells. During substrate phosphorylation, ATP is synthesized during specific enzymatic steps in the catabolism of the organic compound. This ATP is produced by a process called fermentation, which will be discussed in the following section.

GLYCOLYSIS

Glycolysis is one of three phases of the catabolism of glucose to carbon and water process. The other two phases—Krebs cycle and the electron transport system—will be discussed later. Glycolysis can occur under both aerobic and anaerobic conditions. Some of the anaerobic processes are called fermentations. Fermentation is a process whereby the anaerobic decomposition of organic compounds takes place. These organic compounds serve both ultimate electron donors and acceptors. Thus, fermentable substances often yield both oxidizable and reducible metabolites (organic compounds produced by metabolism).

The energy-converting metabolism (fermentation) in which the substrate is metabolized without the involvement of an external oxidizing agent is more easily understood by looking at a metabolic pathway. For example, in some bacteria the fermentation of glucose begins with a pathway called glycolysis.

Glycolysis (sometimes referred to as the Embden-Meyerhof-Parnas pathway—EMP pathway) involves the breakdown or splitting of glucose (sugar) in a catabolic reaction that converts one molecule of glucose into two molecules of the end product, pyruvic acid. In this pathway, energy from energy-yielding (exergonic) reactions is used to phosphorylate ADP—that is, ATP is synthesized from ADP. This is an example of substrate phosphorylation where energy from a chemical reaction is used directly for the synthesis of ATP from ADP.

The end product in the energy-yielding glycolysis process is the release of a small amount of energy that is used for various cell functions and the loss of larger amounts of energy in the form of fermentation products. Common fermentation products of glycolysis include ethanol, lactic acid, alcohols, and gaseous substances produced, for example, by certain bacteria (Singleton, 1992).

Respiration

The process by which a compound is oxidized using oxygen as external electron acceptor is called *respiration*. Using an external electron acceptor is important because in the fermentation process little energy is yielded, mainly because there is only a partial oxidation of the starting compounds in this process. However, if some external terminal acceptor, oxygen for example, is present, all substrate molecules can be oxidized completely to a by-product (carbon dioxide). When this occurs, a far higher yield of ATP is possible.

Because an external oxidizing substance is used, the substrate undergoes a net oxidation (see Figure 8.3). The oxidation of a substrate provides more energy than that obtainable— from the same substrate—as fermentation.

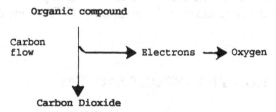

FIGURE 8.3 *Aerobic respiration: the process by which a compound is oxidized using oxygen as external electron acceptor.*

```
Overall Reaction:   Pyruvate + 4NAD + FAD  ⟶  3CO₂ + 4NADH + FADH
                    GDP + Pi  ⟶  GTP
                    GTP + ADP  ⟶  GDP + ATP
                                                              15 ATP
Electron-transport      4 NADH =    12 ATP
phosphorylation           FADH =     2 ATP
```

FIGURE 8.4 *Summary of the overall reaction of the Krebs cycle.*

KREBS CYCLE (TCA)

The Krebs cycle is sometimes called the *citric acid cycle* or *tricarboxylic acid cycle* (TCA), which is commonly called the "energy wheel" of cellular metabolism (see Figure 8.4) because it represents a cyclical sequence of reactions that are crucial to supplying the energy needs of cells.

When oxygen is available to the cell, the energy in pyruvic acid is released through aerobic respiration. In the TCA cycle pyruvate is first decarboxylated (the removal of a carboxyl group from a chemical compound), leading to the production of one molecule of NADH and an acetyl coupled to coenzyme A (acetyl-CoA). The addition of the activated 2-carbon derivative acetyl CoA to the 4-carbon compound oxaloacetic acid forms citric acid, a six-carbon organic acid. The energy of the high-energy acetyl-CoA bond is used to drive the synthesis. After undergoing dehydration, decarboxylation, and oxidation, two additional carbon dioxide molecules are released. Eventually, oxalacetate is regenerated and serves again as an acetyl acceptor, thus completing the cycle. In the course of the cycle, three NADH's, one FADH, and one ATP are produced by substrate phosphorylation. The presence of an electron acceptor in respiration allows for the complete oxidation of glucose to carbon dioxide, with a greater yield of energy (Wistreich & Lechtman, 1980).

ELECTRON TRANSPORT SYSTEM

The electron transport system (ETS) is a common pathway for the utilization of electrons formed during a variety of metabolic reactions. Most molecules are prevented from going into and out of an organism's cells by the cytoplasmic membrane. This is the case, of course, so that the cell can control its internal environment. During metabolism, however, the cell must be able to take in various substrates and get rid of waste. These tasks are accomplished by transport systems. In some organisms—gram-negative bacteria, for example—the transport system is located in membranes other than the cytoplasmic membrane (Singleton, 1992).

A typical ETS is composed of electron carriers. In a bacterium, the ETS involved with respiration occurs in the cytoplasmic membrane. ETS has two functions: (1) to accept electrons from electron donors and transfer them to electron acceptors and (2) to save energy during electron transfer by synthesis of ATP.

The two protein components that form the ETS are the *flavoproteins* and *cytochromes*. *Flavoproteins* are riboflavin-containing proteins (enzymes) that act as dehydrogenation catalysts or hydrogen carriers in a number of biological reactions. The flavin portion, which is bound to a protein, is alternately reduced as it accepts hydrogen atoms, and oxidized when electrons are passed on. Riboflavin, also called vitamin B_2, is a required organic growth factor for some organisms (Singleton & Sainsbury, 1994).

Cytochromes are iron-containing proteins that receive and transfer electrons by the alternate reduction and oxidation of the iron atoms and are important in cell metabolism. Cytochromes in the ETS are distinguished for, among other things, their reduction potentials. One cytochrome can transfer electrons to another that has a more positive reduction potential and can itself accept electrons from cytochromes with a less positive reduction potential (Abeles et al., 1992).

AUTOTROPHIC AND HETEROTROPHIC METABOLISM

Autotrophs can use carbon dioxide as their major carbon source for the formation of essential biochemical compounds. Photosynthetic autotrophic bacteria combine carbon dioxide with ribulose diphosphate to form other macromolecules, which can be used for energy. The nonsynthetic chemosynthetic autotrophs rely on oxidation of inorganic compounds, including hydrogen, for the energy to fix carbon dioxide.

In heterotrophic metabolism, carbon dioxide cannot be used as a major carbon source. Chemosynthetic heterotrophs perform metabolic reactions involving proteins, lipids, and carbohydrates similar to those performed by other organisms. Heterotrophic organisms that are phototrophic can adjust to varying amounts of oxygen. For a review of these concepts and to gain a better understanding of their interrelationships, refer to Chapter 3, Figure 3.3.

MICROBIAL NUTRITION

In earlier chapters, various aspects of the chemical makeup of cell constituents were presented. It is useful at this point to summarize what is known about the chemical composition of a bacterial cell (Table 8.1).

From Table 8.1 it is clear that cells contain large amounts of water, inorganic, and organic molecules, but consist primarily of macromolecules such as proteins and nucleic acids. The cell is

Table 8.1. Chemical composition of a bacterial cell.

Molecule	Percent Wet Weight	Percent Dry Weight
Water	70	—
Total macromolecules	26	96
Proteins	15	55
Polysaccharide	3	5
Lipid	2	9
DNA	1	3
RNA	5	20
Total monomers	3	0.5
Amino acids	0.5	0.5
Sugar	2	2
Nucleotides	0.5	0.5
Inorganic ions	1	1
Totals	100%	100%

Source: Data compiled from Neidhart (1987).

capable of obtaining most of the water (small molecules) it needs from the environment in usable form, whereas macromolecules are synthesized inside the cell.

The mass of the cell primarily consists of four types of atoms: carbon, oxygen, nitrogen, and hydrogen. A number of other atoms are functionally important to the cell but are less apparent. Calcium, magnesium, iron, zinc, and phosphorus are present in microbial cells but in lesser amounts than carbon, hydrogen, oxygen, and nitrogen.

NUTRITION

Nutrients used by organisms and obtained from the environment can be divided into two classes: (1) macronutrients, which are required in large quantities, and (2) micronutrients, which are required in lesser quantities.

Macronutrients

Most procaryotes require an organic compound to obtain their source of carbon. Bacteria have demonstrated that they can assimilate a wide variety of organic carbon compounds to make new cell material. Major macronutrients such as amino acids, fatty acids, organic acids, sugars, and others are used by a variety of bacteria. The major macronutrients in the cell, after carbon, are nitrogen, sulfur, phosphorus, potassium, magnesium, calcium, sodium, and iron. Table 8.2 shows some of the common forms of these major elements needed for biosynthesis of cell components.

Micronutrients

Micronutrients, or trace elements, are also required and just as critical to the overall nutrition of a microorganism as are macronutrients. For example, cobalt is needed for the formation of vitamin B_2; zinc plays a role in the structure of many enzymes; molybdenum is important for nitrate reduction; and copper is important in enzymes involved in respiration.

Table 8.2. Macronutrients.

Elemental Forms Found in the Environment	Element
Carbon dioxide	Carbon
Organic compounds	
Water	Hydrogen
Organic compounds	
Water	Oxygen
Oxygen gas	
Ammonia	Nitrogen
Nitrate	
Amino acids	
Phosphate	Phosphate
Hydrogen sulfide	Sulfur
Sulfate	
Organic compounds	

SUMMARY OF KEY TERMS

- *Metabolism:* the complex of physical and chemical processes involved in the maintenance of life
- *Phosphorylation:* the changing of an organic substance into organic phosphate
- *Electron Transport System:* a series of electron carriers that operate together to transfer electrons from donors such as NADH and FADH$_2$ to acceptors such as oxygen
- *Glycolysis:* the anaerobic conversion of glucose to lactic acid by use of the Embden-Meyerhof-Parnas pathway
- *Krebs cycle:* the cycle that oxidizes acetyl coenzyme A to carbon dioxide and generates NADH and FADH$_2$ for oxidation in the electron transport system

CHAPTER 9

Microbial Growth

BECAUSE their primary function is to prevent the spread of waterborne disease, water and wastewater treatment specialists have a keen interest in the control of microbial growth. On the one hand, the water specialist's primary concern has to do with preventing pathogenic microorganisms from entering the potable water distribution system. The wastewater specialist, on the other hand (while also concerned with preventing pathogenic microorganisms from being part of the effluent dumped into a receiving body—lake, pond, stream, or river), is mainly concerned with providing optimal growth conditions for microorganisms used to stabilize wastes in biological treatment processes. Control of microbial growth requires a fundamental knowledge of microbial growth activities.

Unlike industrial applications where pure cultures of microorganisms are used, water and wastewater treatment use the mixed (batch) culture commonly found in waste and in the natural environment (Sundstrom & Klei, 1979). All forms of microorganisms will grow and multiply when ample nutrients are present. The type of nutrient, source of energy, availability of water, appropriate temperature, appropriate pH, appropriate levels (or the absence of) oxygen, specific metabolic rate, and its position in the food chain controls the predominance of any microorganism. Water and wastewater specialists are concerned about the presence of a variety of microorganisms in the water and in the waste stream. However, their focus may be on the presence of a particular type of organism or, as in the case of wastewater treatment, their focus may be directed toward the

117

presence and destruction and/or maintenance of a mass or biomass concentration. Since most information is available about bacteria and control of its growth, most of the discussion in this chapter concerns bacteria.

BACTERIAL GROWTH

In microbiology, *growth* may be defined as an increase in the number of cells or in cellular constituents. If the microorganism is a multinucleate (coenocytic) organism in which nuclear division is not accompanied by actual cell division (as with bacteria), growth results only in an increase in cell size and not in cell number. In bacteria, as a general rule, growth leads to a rise in cell number because reproduction is by *binary fission* where two cells enlarge and divide into two progeny of about equal size. Other bacterial species increase their cell numbers asexually by *budding*, for example, as with the mycoplasma (Wistreich & Lechtman, 1980).

Population Growth

It is not usually convenient or practical for the water or wastewater specialist to investigate the growth of individual microorganisms because of their small size. Therefore, plant operators normally follow changes in the total population number when studying growth.

As was mentioned earlier, growth is defined as an increase in the number of microbial cells in a population, which is measured as an increase in microbial mass. The change in cell number or mass per unit time is the *growth rate*.

When bacterial cells are introduced into a suitable medium, held at the optimum growth temperature, and a small volume of medium is withdrawn and cultured (bacteria growing in or on a medium), a count can be made of the cells it contains (counting methods will be discussed later). In this way the development of a population (i.e., the increase in cell numbers with time) can be observed and followed. By plotting the number of cells against time, a *growth curve* can be obtained. The actual shape of each portion of the curve and the actual numbers of organisms obtained vary according to species and type of media used (Singleton, 1992).

BACTERIAL GROWTH CURVE

The growth of bacteria can be plotted as the logarithm of cell number versus the incubation time. The resulting curve has four distinct phases. It must be pointed out that these phases reflect the events in a population of bacteria or other microorganisms, not the individual cells. The terms *lag, log, stationary,* and *death phase* do not apply to individual cells but only to populations of cells (see Figure 9.1).

Lag Phase of Growth

When bacteria are inoculated into a culture medium, growth usually does not begin immediately. Apparently, this lag in growth, which may be brief or extended, represents a transition period on the part of bacteria transferred to new conditions. During this transition period, time is required for acclimation

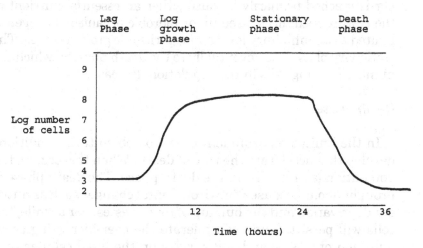

FIGURE 9.1 *Bacterial growth curve. The four phases of the growth curve are identified on the curve and discussed in the text. [Adapted from McKinney (1962), p. 118.]*

and for performing various functions such as synthesis of new enzymes. When conditions are suitable, binary fission begins, and after an acceleration in the rate of growth, the cells enter the logarithmic (log) phase.

The Logarithmic Phase

During the log phase, the number of bacteria are growing at the maximal rate possible. The bacteria increase in a geometric progression—one splits to make two, two split to make four, four to eight, and so on. Because each individual divides at a slightly different moment, the growth curve rises smoothly (in a straight line) rather than in discrete jumps. From this logarithmic increase in cell number, we can calculate the average time for a cell to divide, the *generation time.*

Stationary Growth Phase

Eventually, population growth ceases and the growth curve becomes horizontal (Figure 9.1). When this occurs, the growth and death rates are more nearly identical, and a fairly constant population of bacteria is achieved. This uniformity in population number is reached primarily because either an essential nutrient of the culture medium is used up or (aerobic organisms) oxygen is limited to an inhibitory level and logarithmic growth ceases. The stationary phase leads eventually to the death phase in which the number of living cells in the population decreases.

Death Phase

In the limited environment of the batch culture, conditions develop that accelerate the rate of death. When this occurs, the population is said to be in the death phase. The death phase is brought about because of environmental changes such as nutrient deprivation and the buildup of toxic wastes. For a while, the cells will persist. Some will tolerate the ever-increasing accumulation of wastes and will survive on the lysed cellular contents of the dead cells. At some point, however, further degradation of conditions will cause even the hardiest organism to die.

CONTINUOUS CULTURE

The batch processes for growing bacteria are often adequate for various biochemical studies or in plant analyses conducted by water and wastewater specialists. However, since there is usually considerable variation in the ages of the resulting cells and their metabolic activities, the *continuous culture technique* is used to avoid these variations. Continuous culture allows cells to grow under constant, controlled, defined conditions. Two major types of continuous culture systems are currently in use: *chemostats* and *turbidostats*. A chemostat permits control of both the population density and the growth rate of the culture. The turbidostat has a photocell that measures the absorbance (turbidity) of the culture in the growth vessel.

THE EFFECT OF ENVIRONMENTAL
FACTORS ON GROWTH

The growth of microorganisms is greatly affected by the chemical and physical conditions of their environments. An understanding of environmental influences helps in the control of microbial growth and in the understanding of the ecological distribution of microorganisms.

Temperature

Temperature is one of the most important environmental factors affecting the growth and survival of microorganisms. In turn, one of the most important factors influencing the effect of temperature upon growth is *temperature sensitivity* of enzyme-catalyst reactions. As temperature rises, enzyme reactions in the cell proceed at more rapid rates (along with increased metabolic activity) and the microorganism grows faster. However, above a certain temperature, growth slows. Eventually, as the temperature continues to increase, enzymes and other proteins are denatured and the microbial membrane is disrupted. Thus, the microorganism is damaged or killed off. Usually, therefore, as the temperature is increased, functional enzymes operate more rapidly up to a point where the microorganisms may be damaged and inactivation reactions set in.

Because of these opposing temperature influences, for every organism there is a minimum temperature below which growth no longer occurs, an optimum temperature at which growth is most rapid, and a maximum temperature above which growth is not possible (see Figure 9.2). The temperature optimum is always nearer the maximum than the minimum. Although these three temperatures, called the *cardinal temperatures*, are generally characteristic for each type organism, they are not rigidly fixed but often depend to some extent on other environmental factors such as available nutrients and pH.

The cardinal temperatures of different microorganisms differ widely (Table 9.1). Some microbes have temperature optima ranging from 0°C to as high as 75°C. The temperature range through which growth occurs is even wider, from below freezing to greater than boiling. Generally, the growth temperature range for a particular microbe is about 30–40 degrees.

Microorganisms occupying those groups listed in Table 9.1 can be placed in one of four other groups based on temperature range for growth.

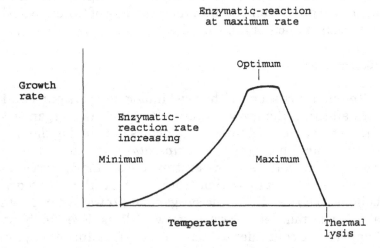

FIGURE 9.2 *Effect of temperature on growth rate and the enzymatic-reaction activity that occurs as the temperature increases.*

Table 9.1. Approximate temperature ranges for microbial growth.

Microorganism	Temperature Ranges (°C)		
	Minimum	Optimum	Maximum
Bacteria (nonphotosynthetic)	−10 to 85	10 to 105	25 to 110
Bacteria (photosynthetic)	70	30 to 80	45 to 85
Eucaryotic algae	−40 to 35	0 to 50	5 to 57
Fungi	0 to 25	5 to 50	15 to 60
Protozoa	2 to 29	20 to 45	31 to 49

Source: Adapted from Prescott et al (1993), p 126.

1. *Psychrophiles* (with low-temperature optima) grow optimally at or below 15°C, do not grow above 20°C, and have a lower limit for growth of 0°C or below. These microorganisms are readily isolated from polar seas. Psychrophilic microorganisms have adapted to their cold environments in several ways. For example, their transport systems, protein synthetic mechanisms, and enzymes function at low temperature. Because psychrophiles are found in environments that are constantly cold, and they are rapidly killed by warming to room temperature, their laboratory study is difficult. Psychrophilic microorganisms include a number of gram-negative bacteria, certain fungi and algae, and a few gram-positive bacteria.

2. *Psychrotrophic* bacteria (with low-enzyme and low-temperature optima) can grow at a low temperature range of 0–5°C, but they grow optimally above 15°C, with an upper limit for growth range of 25–30°C, with maximum about 35°C. Psychrotrophic fungi and bacteria are major contributors to spoilage of refrigerated food such as meat, milk, vegetables, and fruits; only when these foods are frozen is microbial activity (growth) not possible.

3. *Mesophilic* bacteria (with mid-range-temperature optima) grow optimally at temperatures of 20–45°C. Most microorganisms fall within this category, and they include bacteria that are pathogens in man and other animals.

4. *Thermophilic* bacteria (with high-temperature optima) can

grow at temperatures of 55°C or higher. Their growth minimum is around 45°C, and they often have optima of 55–65°C.
A few thermophiles have maxima above 100°C. These
thermophiles occur in composts, hydrothermal vents on the
ocean floor, and in hot springs (Singleton & Sainsbury,
1994).

pH

When attempting to control microbial growth, controlling pH
is one of the best methods (McKinney, 1962). Most bacteria
grow best at or near neutral pH 7, and the majority cannot grow
under either acidic or alkaline conditions. Acidity or alkalinity
of a solution is expressed by its *pH value.* pH value is defined as
the hydrogen ion activity of a solution, that is, pH is the negative logarithm of the hydrogen ion concentration. This hydrogen concentration is important because it affects the equilibrium relationship of many biological systems that function only
in a very narrow pH range (Tchobanoglous & Schroeder, 1987).
pH can be expressed as follows:

$$pH = -\log (H^+) = (1 / H^+)$$

The pH scale extends from pH 0–pH 14. pH values less than 7
are acidic, and those greater than 7 are alkaline (basic or caustic). Each pH unit represents a tenfold change in hydrogen ion
concentration. Therefore, vinegar with a pH near 2 and ammo-
nia (household-variety) with a pH of 11 differ in hydrogen ion
concentration by one billion times. The following points allow
for a better understanding of the hydrogen ion–pH relationship:

1. When the hydrogen ion concentration is highest, pH is lowest (solution is very acidic).
2. When the hydrogen ion concentration is lowest, pH is the
   highest (solution is most alkaline).
3. pH 7 is neutrality (midpoint on the scale). pH values lower
   than 7 represent increased ion concentration (more acidic)
   than neutral point, and pH values higher than 7 represent

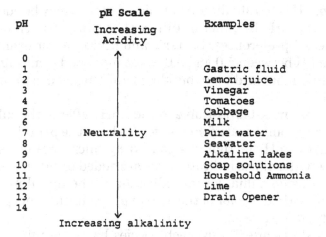

FIGURE 9.3 *pH scale.*

lower hydrogen ion concentrations (more alkaline) than neutral point.

From Figure 9.3 it is apparent that the habitats in which microorganisms may live can vary widely—from pH 2 at the acid end to alkaline-solution soils that may have pH values around 10.

As previously stated, pH dramatically affects microbial growth. Each species has a definite pH growth range and pH growth optimum that can be clearly seen in Table 9.2.

In culturing microorganisms, pH adjustment of the culture medium is a common practice. For example, if the pH is too acidic, an alkaline substance, such as sodium hydroxide (caustic solution), may be added. On the other hand, if the culture me-

Table 9.2. Approximate effect of pH on microbial growth.

Microorganism Type	Lower-Limit Range	Upper-Limit Range
Bacteria	0.5	9.5
Algae	0.0	9.9
Fungi	0.0	7.0
Protozoa	3.2	9.0

Source: Adapted from Prescott et al. (1993), p. 125.

dium pH is too alkaline, an acidic substance may be added to adjust the pH. In general, different groups of microorganisms have pH preferences. Bacteria and protozoa, for example, prefer a pH between 5.0 and 8.0. Most fungi and some algae prefer slightly more acidic surroundings, ranging from a pH of about 3 to 6.

Sometimes it is desirable to add a pH *buffer* to the culture medium to compensate for changes (to keep the pH relatively constant) in pH fluctuations caused by microorganisms as they grow. Oftentimes indicator dyes are added to culture media and can provide a dual visual indication of the initial pH and also changes (in the color of the dye) in pH resulting from growth activity of microbes.

As stated previously, each species has a definite pH growth range and pH growth optimum. This important characteristic gives the water and wastewater specialist significant control of various microbial populations. Most natural environments have pH values between 5 and 9; organisms in this optima range are most common. Microorganisms that live at low pH are called *acidophiles*. Microorganisms with a high optima, ranging from 8.5–11.5, are called *alkalophiles*. It is interesting to note that acidophiles and alkalophiles may not grow at all or only very slowly at pH 7 (neutral pH).

Water Availability

The availability of water is another environmental factor that can affect the growth of microbes. In some microbes (e.g., bacteria) approximately 80% or more of their mass is water. In the growth phase, nutrients and water products enter and leave the cell, respectively, in solution. The point is that for these microbes to grow, they must be in or on an environment that has adequate available water or ions in solution.

All microorganisms (like all other organisms) need water for life. Indeed, availability of water is critical. Water availability depends not only on water content of the environment but also on the substances that are present; that is, some substances can absorb water and do not readily give it up.

Water availability is generally expressed in physical terms as *water activity* (a_w). Singleton and Sainsbury (1994) define water activity as the "amount of free or available water in a given substance" (p. 950). The water activity of a solution is 1/100 the relative humidity of the solution when expressed as a percent. Water availability is expressed as a ratio of the vapor pressure of the air over the solution divided by the vapor pressure at the same temperature of pure water as demonstrated in the following:

$$a_w = \frac{p_{\text{solution}}}{p_{\text{water}}}$$

Values of water activity vary between 0 and 1. Some representative values are given in Table 9.3.

Water diffuses from a region of high water–low solute concentration to a region of lower water–higher solute concentration. Hence, if pure water and a salt solution are separated by a semipermeable membrane, water will diffuse from the pure water into the salt solution by *osmosis*. The cytoplasm of most cells has a higher solute concentration than the environment; therefore, water diffuses into the cell. On the other hand, if the cell is in an environment of low water activity, water will flow out of the cell. An environment such as salt or sugar in solution has low water activity and causes the cell to give up water. When this occurs, the plasma membrane shrinks away from the wall (plasmolysis) and the cell dehydrates, damaging the membrane. The cell ceases to grow.

Table 9.3. Water activity of various materials.

Material	Water Activity Value
Pure water	1.000
Human blood	0.995
Bread	0.950
Ham	0.900
Jams	0.800
Candy	0.700

Oxygen

Microorganisms vary in their need for, or tolerance of, oxygen. Some bacteria, for example, need oxygen for growth. Others need the absence of oxygen for growth, and still others can grow regardless of the presence or absence of oxygen. Microorganisms can be grouped depending on the effect of oxygen on growth. A microorganism that is able to grow in the presence of atmospheric oxygen is an *aerobe*, whereas one that can grow in its absence is an *anaerobe*. Organisms that depend on atmospheric oxygen for growth are called *obligate aerobes*. Microorganisms that do not require oxygen for growth, but grow better in its presence, are called *facultative anaerobes*. The strict or obligate anaerobes, on the other hand, do not tolerate oxygen and will die in its presence.

In order to grow some aerobes, it is necessary to aerate. This is the case because oxygen is poorly soluble in water, and the oxygen that is used up by microorganisms during growth is diffused from air too slowly. To compensate for this shortage of oxygen in cultured aerobes, forced aeration is desirable. This can be accomplished by forcing sterilized air into the medium or by simply shaking the tube or flask vigorously.

The anaerobic culture requires the exclusion of oxygen. This is a difficult task; oxygen, obviously, is readily available in the air. In order to vacate air from the culture media in which anaerobes are to be grown, it is necessary to completely fill tubes to the top with culture medium and seal with tight-fitting stoppers. This procedure is useful for providing anaerobic conditions for organisms that are not too sensitive to small amounts of oxygen. It is also possible to add a reducing agent to the medium, which reacts with oxygen and excludes it from the culture medium. A common reducing agent used in this procedure is thioglycollate. The nature of bacterial oxygen responses can be readily determined by growing bacteria in culture tubes filled with a solid culture medium or media treated with a reducing agent (see Figure 9.4). In order to detect the presence of oxygen easily in the medium, an indicator dye is usually added; the dye will change color and indicate the penetration level of oxygen in the medium.

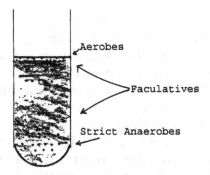

FIGURE 9.4 *Oxygen needs of microorganisms as revealed by the position of microbial colonies within the tube of culture medium. Aerobes grow only near the surface. Facultatives grow throughout the tube. Strict anaerobes grow only in the bottom of the tube where there is no oxygen.*

Ultraviolet Radiation

Except for photosynthetic bacteria, ultraviolet radiation (UV) can be used to kill all kinds of microorganisms. As an alternative to chlorination for disinfection, ultraviolet radiation is gaining acceptance in water and wastewater treatment methodologies.

While water and wastewater disinfection by chlorination is still widely used in the United States, various regulatory agencies are taking a close look at chlorine and the hazards associated with its use; they are proposing regulations to control chlorine use as a disinfectant. This is not surprising, however, especially since chlorine use has several disadvantages: worker safety concerns, carcinogenic by-product production, inability to destroy or inactivate certain pathogens, and excessive toxicity to aquatic life. These problems are avoided with UV radiation. When installed, UV radiation is generally placed in dense networks of lamps to ensure that the water stream receives adequate exposure (blanket effect) to UV radiation. This technique provides an additional advantage over chlorination because UV radiation allows for less contact time. UV radiation kills (alters the genetic material in cells of microorganisms) many microorganisms owing to its short wavelength (10–400 nm) and high energy content. The most lethal UV has a wavelength of approximately 260 nm. This wavelength is most effectively absorbed by DNA, which in turn, is damaged.

UV radiation used for disinfection has another advantage over chemical treatment processes such as chlorination. For example, chlorine is a hazardous material that requires specialized training for routine chlorine cylinder change-out procedures, and training plant personnel to Hazardous Materials Emergency Responder Levels I–III is time consuming and expensive. Voutchkov (1995) reports that the cost-benefit comparisons of "UV and chlorine disinfection systems are comparable" (p. 40). Along with the advantages of using UV radiation for disinfection, several disadvantages must be taken into account. The two major disadvantages of using UV in wastewater treatment are (1) UV radiation has poor penetrating ability, especially in water or waste streams that have high turbidity, and (2) UV radiation systems can require a large amount and high degree of labor-intensive maintenance.

SUMMARY OF KEY TERMS

- *Bacterial Growth Curve:* when microorganisms are cultivated in a batch culture (closed system) and incubated, their growth can be plotted on a graph as the logarithm of cell numbers versus incubation time. The resulting curve has four distinct phases.
  1. *Lag phase:* period following inoculation of microorganisms into fresh culture media when there is no increase in cell numbers or mass
  2. *Log phase* (or exponential phase): phase of growth curve during which the microbial population is growing at a constant and maximum rate, dividing and doubling at regular intervals
  3. *Stationary phase:* phase of microbial growth when population growth ceases and the growth curve levels off
  4. *Death phase:* phase of microbial growth when the viable microbial population declines

CHAPTER 10

Pathogenicity

I can think of a few microorganisms, possibly the tubercle bacillus, the syphilis spirochete, and malarial parasite, and a few others, that have a selective advantage in their ability to infect human beings, but there is nothing to be gained, in an evolutionary sense, by the capacity to cause illness or death. Pathogenicity may be something of a disadvantage for most microbes, carrying lethal risks more frightening to them than to us. The man who catches a meningococcus is in considerably less danger for his life, even without chemotherapy, than meningococci with the bad luck to catch a man. Most meningococci have the sense to stay out on the surface, in the rhinopharynx. During epidemics this is where they are to be found in the majority of the host population, and it generally goes well. It is only in the unaccountable minority, the "cases," that the line is crossed, and then there is the devil to pay on both sides, but most of all for the meningococci. (Thomas, 1974, pp. 76–77)

NOTWITHSTANDING Thomas's account of the dilemma the meningococci confront upon entering the human body, water and wastewater treatment specialists are most concerned with preventing an invasion by waterborne pathogens into human beings and other forms of life. The most frequent cause of waterborne disease in public water supplies in the United States is inadequate water treatment; that is, either treatment is nonexistent or is ineffective owing to treatment-process breakdown, especially in disinfection. When inadequate water treatment or treatment-system breakdown occurs, most diseases that are transmitted via the untreated water are caused

131

by contaminated fecal material. Table 10.1 shows the public water system breakdowns in the United States reported by the Centers for Disease Control (CDC) during the 1986–1988 time frame.

Even before Mary Mallon (a.k.a. Typhoid Mary) was "cooking up" her daily food preparations (she was an American cook in the late 1800s and early 1900s) and passing on typhoid to her unsuspecting victims, people had suspected a relationship between microorganisms and disease. As a matter of historical record, even before the discovery of waterborne, disease-causing microorganisms, people suspected a relationship between water and the spread of disease. The quest for the truth about disease and disease-causing microorganisms began almost 2,600 years ago. This was the time frame when Hippocrates, the father of medicine, suspected that "different" waters caused several different diseases. It took several more centuries before Leeuwenhoek invented the microscope in 1675. Leeuwenhoek's microscope, with several later refinements, opened up the microscopic world. *Giardia lamblia* was the first microorganism studied and described by Leeuwenhoek. Eventually, in the nineteenth century, other great men of science such as Koch and Pasteur disproved the ancient theory of poisonous vapors arising from decaying filth as the cause of disease; instead, they developed the germ theory.

At the beginning of this chapter, Lewis Thomas discussed the plight of the meningococci that might have the misfortune of

Table 10.1. Water system breakdowns leading to waterborne disease outbreaks (1986–1988).

Type of Breakdown or Deficiency	Community Water Systems	Noncommunity Water Systems	Total
Untreated surface water	1	1	2
Untreated ground water	3	9	12
Treatment	11	12	23
Distribution system	6	0	6
Miscellaneous	1	2	3
Total	22	24	46

Source: Centers for Disease Control, Atlanta.

entering a human body. After reading Thomas's discussion, the reader may have developed the misperception that pathogenic microorganisms are defenseless and not at home within the human body. This is not the case. Microorganisms can adapt. Consider the following descriptive account of the adaptive challenges faced by waterborne pathogenic bacteria when they are ingested.

> In water, this bacterium was experiencing an environment where the temperature was well below 37 degrees C, nutrient concentrations and osmotic strength were low, and pH was near neutral. At least some oxygen was available. When the bacterium is ingested, it suddenly encounters a higher temperature, higher osmotic strength, a transient exposure to low pH in the stomach, followed by a rise in pH in the intestine and high concentrations of membrane-disrupting bile salts. Also, the environment of the small intestine, and to a greater extent the colon, is anaerobic. The bacterium will find abundant sources of carbon and energy in the intestine, but the forms of these compounds will be different from those it may have been using in water. Keep in mind that the incoming bacterium does not have much time to adapt to its new environment because it takes only a few hours to transit the small intestine and reach the colon, where it also will encounter stiff competition from the resident microflora. (Salyers & Whitt, 1994, pp. 64–65)

Given the adaptive challenges faced by the waterborne bacteria and their ability to make these adaptations, it is not surprising that pathogenic bacteria have evolved defense mechanisms that make their destruction increasingly more difficult. This is the ultimate challenge confronting water and wastewater treatment specialists of the future. Given the fact that pathogenic microorganisms have the remarkable ability to adapt and to survive, it follows that advanced biological treatment methodologies will have to evolve and to adapt as well.

CAUSAL FACTORS FOR TRANSMISSION OF DISEASE

Certain factors must exist for the transmission of disease. These factors are related to the diseased individual (the host), the microorganism (the agent), and the environment. Among the many diseases of microbial origin, some are caused by viruses, some by fungi, some by bacteria, and others by protozoa. It is important to note that disease does not necessarily follow exposure to a given pathogen. For disease to develop, several factors must be present.

1. Pathogen (causative agent): any microorganism that can cause disease is called a *pathogen*. Normally this pathogen is parasitic and lives on or in another organism. In some diseases the link between pathogen and disease is very specific; that is, the pathogen may infect only certain species. Along with being selectively specific, for a pathogen to cause the occurrence of disease depends on various factors. These factors include the extent or degree of resistance of the host and the capacity to produce disease (*virulence*) of the pathogen (Singleton, 1992).

2. Pathogen's reservoir: the pathogen lives and reproduces (multiplies) in a reservoir and is unable to reproduce or grow outside the reservoir. Humans, animals, plants, organic matter, and/or soils can serve as the pathogen's reservoir. The human body is a significant reservoir of microbial pathogens; these organisms are a normal part of human microflora and usually do not behave as pathogens unless disturbed in surgery or in an injury. When this occurs, specific or nonspecific defenses have been impaired; the type of pathogenic contamination that takes place is known as *opportunistic* (Salyers & Whitt, 1994).

3. Movement of pathogens from reservoir: movement of pathogens in humans may be from body openings in the respiratory, urinary, and intestinal systems, from open infections, and by mechanical means delivered by disruptions or damage caused by injuries.

4. Transmission of disease to another host: transmission can be accomplished directly or indirectly. In direct transmission disease passes via an animal bite, for example, immediately to a new host. Indirect transmission is accomplished through vectors or vehicles. These vectors can be mosquitos, ticks, fleas, and other invertebrates. The vehicles are sub-

stances such as milk, food, air, and other nonliving substances.

5. Virulence of the pathogen: the occurrence of disease depends upon the pathogen's virulence (its capacity to cause disease).

6. Degree of resistance to transmission: humans possess mechanisms of defense against disease. For example, human skin provides the primary defense against infectious disease. In order for most diseases to pass or be transmitted through the skin's resistance barrier, skin must be broken by wounding or insect bites as explained in factor number 4.

For disease to spread there must be some vehicle or medium or opportunity for spread. This is where water and wastewater treatment specialists play such a vital role. If untreated water is ingested by an individual, the individual may become a carrier of waterborne disease—as in the case of Typhoid Mary. Mary infected several people with typhoid fever, but she never demonstrated obvious symptoms of this deadly disease; thus, she was a carrier.

Pathogenic organisms found in water and wastewater may be from carriers of disease. The principal categories of pathogenic organisms found in water and wastewater are bacteria, viruses, protozoa, and helminths. Such highly infectious organisms have been responsible for countless numbers of deaths and continue to cause the death of large numbers of people and animals in underdeveloped parts of the world.

Problems related to waterborne diseases in underdeveloped parts of the world are well documented. For example, Ries et al. (1992) report that in 1991, Piura, a Peruvian city of more than 350,000, recorded that within a 2-month period more than 7,922 cases and 17 deaths were attributed to a cholera epidemic. During the investigation, a hospital-based culture survey showed that approximately 80% of diarrhea cases were cholera. A study of 50 case-patients and 100 matched controls demonstrated that cholera was associated with drinking unboiled water, drinking beverages from street vendors, and eating food from vendors. It is interesting to note that in a second study, patients were more likely than controls to consume beverages with ice. Ice was produced from municipal water. Testing of municipal water supplies revealed none or insufficient chlorination, and fecal coliform bacteria were detected in most samples tested. With epidemic cholera spreading throughout Latin America, these findings emphasize the importance of safe municipal drinking water. More specifically, the results of this study indicate that even though we have come a long way in the fight against waterborne disease, some parts of the world are still struggling with this life-threatening problem.

PARASITES AND PATHOGENS

One form of pathogenicity is demonstrated when parasitic microorganisms derive their nutrients for growth and reproduction from living on or in a viable host. Another form of pathogenicity occurs when toxic substances are produced by pathogens (McKinney, 1962).

Microbial parasites include viruses, protozoans, and helminths—while bacteria, fungi, and actinomycetes are pathogens—all of which are of concern to water and wastewater specialists. *All* viruses are parasitic. Protozoans are parasites that commonly enter and live in the gastrointestinal system of humans and animals. Two parasitic forms of helminths (worms) are of concern to water and wastewater specialists: roundworms and flatworms (tapeworms).

In water treatment, parasites and pathogens are treated with chemicals and removed through the filtering methods recommended in *Standard Methods for the Examination of Water and Wastewater* (American Public Health Association, 1989). The removal of parasites and pathogens in wastewater treatment is a complex process. A better understanding of this removal process can be gained by following the movement of the waste stream, including parasites and pathogens, through the basic wastewater treatment process.

When influent leaves pretreatment processes such as screening and comminuting, no major reduction in parasites and pathogens has yet been accomplished. While the waste stream is in primary clarification, some removal occurs, but it depends on detention time; that is, as detention time increases the rate of removal increases. In this phase, better removal is possible for the larger microorganisms such as tapeworm eggs and some protozoan cysts. Solid wastes generated in the primary process usually retain a high parasite and pathogen population. In the activated sludge or trickling filter stage, removal is increased for viral organisms. In secondary clarification, sedimentation increases the rate of reduction in helminth eggs, protozoan cysts, some pathogens and large populations of parasites. In anaerobic digestion, parasites are destroyed in large numbers—where destruction is a function of time and temperature; when the time or temperature is increased, destruction increases. When dewatered sludge (biosolids) is incinerated, the destructive kill rate is as absolute as can be attained for the destruction of parasites and pathogenic microorganisms.

Along with the physical and biological processes just described, chemical processes can be utilized to treat water and wastewater streams for parasites and pathogens. For example, chemicals can be used to adjust pH level. Increasing pH level to pH 12 destroys or inactivates many parasites and pathogens. However, it must be noted that when pH drops, some of the inactivated parasites and pathogens can become active; thus, lower pH levels may lead to infection and infestation.

From the preceding it should be clear that like their ability to adapt and survive in the human body, parasites and pathogens are not defenseless against the efforts to destroy or remove them from the water or wastewater stream. This ability of pathogenic microorganisms to adjust or adapt should come as no surprise since humans, through evolution, have also been able to develop ways in which to protect themselves (build up resistance) from parasites and bacteria. The major point to remember is that during the human evolutionary process, pathogenic microorganisms have not remained static. On the contrary, as stated previously, pathogenic microorganisms have also evolved. As a case in point, consider the effect that certain antibiotics have today in relation to the same pathogens that they were used successfully against years ago. Today, these same antibiotics may not be effective because pathogens have evolved ways in which to circumvent the effect of antibiotics to the point where they may no longer provide a defense against disease. Salyers and Whitt (1994) point out the striking way in which pathogenic bacteria have been able to develop several strategies against the various human defensive actions. For example, bacterial cellular structures such as capsule formation, thickened cell walls, and spore formation are defensive mechanisms bacteria have developed with time and constant exposure to various antibiotics. "Bacteria are clearly Machiavellian" (p. 2).

Bacteria are not alone in their resistance to preventing their own destruction. Take for example the viruses. They are protected from most chemical treatment processes because of the very nature of their own chemical composition (McKinney, 1962). Protozoans are other examples of adaptation and adjustment. For instance, they have developed a specialized hyaline wall structure that protects its cyst forms. Moreover, the helminths are often able to resist destruction because of the tough shell that surrounds their eggs.

In wastewater work, disease can be transmitted only when the noninfected individual comes into contact with or ingests disease-carrying wastewater, sludge, or aerosols. For example, the ingestion of wastewater or sludges may permit *Salmonella typhi* to infect the host, possibly leading to typhoid fever. Several other pathogenic bacteria such as *Escherichia coli*, *Vibrio cholerae* (cholera), and *Yersinia* spp. (plague) are usually found in large numbers in untreated wastewater and sludge. The major source of pathogenic bacteria is from human feces, which make up approximately 34% of feces. The actual bacterial and organic composition of wastewater, however, is different from feces. (The actual presence and concentration of pathogenic bacteria in wastewater depends on the contributing population's level of infection.)

Parasites and bacteria are not the only pathogenic microorganisms present in the water and/or wastewater stream. Pathogenic actinomycetes (often considered as a separate group of organisms because they are sometimes listed as bacteria and at other times as fungi) such as *Actinomyces israelii*, *Mycobacterium tuberculosis* (the major pathogen that leads to TB), *Nocardia asteriodes*, and *Nocardia brasiliensis* are of major concern. Diseases transmitted by the actinomycetes group in wastewater are a result of exposure to contaminated aerosols and ingestion of contaminated materials. For example, inhalation of aerosols contaminated with *Nocardia asteriodes* can lead to pulmonary infection.

The pathogenic fungi that are normally present and always a concern in thermophilic composting operations using biosolids (treated sludge) is the *Aspergillus fumigatus*. This pathogen grows on decaying vegetation and can cause significant mycoses of skin, hair, and nails and, although rare, lead to respiratory infection (*Mosby's Medical, Nursing, and Allied Health Dictionary,* 1990). Another pathogenic fungi of concern to wastewater operators is *Candida albicans*, which can cause localized infections of the mouth and skin. Fungi are strict aerobes; thus they can only live in places with close proximity to air (e.g., skin).

Parasitic protozoans can survive outside the intestinal tract in feces as a cyst with a thick hyaline wall that provides resistance to water and wastewater treatment. The protozoan cyst population in wastewater and biosolids depends on the degree of disease prevalence in human and animal contributors. The parasitic protozoans of concern are *Balantidium coli* and *Giardia lamblia,* which is of highest concern and causes giardiasis. *Giardia* infests the small intestine and can lead to severe diarrhea (traveler's diarrhea) and weight loss.

The highest prevalence of giardiasis in the United States is in communities using surface water supplies where potable water treatment consists primarily of disinfection. Giardiasis can also be contracted from contaminated food, mountain streams, and wastewater contaminated with infected feces. The best control against protozoan waterborne infection is proper sanitation procedures (i.e., boil water and wash hands with soap and water).

When sampling for *Giardia lamblia* in water treatment, routine coliform counts (to be discussed in detail later) are inadequate for evaluation of presence and number of cysts because they are extremely resistant to chlorination—especially at generally accepted chlorine residual values. A filtration method for high volume sampling (HVS) can be employed as indicated in the current edition of *Standard Methods.*

When sampling for *Giardia lamblia* in wastewater treatment, biosolids or sludges should be examined; this is where the cysts are the most concentrated. Further sampling instructions can be obtained from a standard parasitology laboratory manual or from the current editions of *Standard Methods*.

The final group of pathogenic organisms of concern to the wastewater treatment plant operator is the helminths. Various pathogenic roundworms (nematodes), such as *Ascaris lumbricoides*, and tapeworms (cestodes), such as *Taenia solium*, are found in wastewater.

SUMMARY

After having discussed the large variety of pathogenic microorganisms and some of their elaborate defense mechanisms against treatment, the obvious question becomes: Will these organisms survive the types of treatment processes that are being used in water and wastewater treatment today? The jury is still out on the definitive answer to this question. Perhaps the best that can be said, at this point in time, is that the longer the time period pathogenic microorganisms are exposed to treatment, the better the chance of destroying or inactivating them.

SUMMARY OF KEY TERMS

- *Pathogenicity:* the ability of an organism to produce disease
- *Virulence:* the capacity to produce disease; a function of microbial invasiveness and toxigenicity measured with reference to a particular host
- *Opportunistic:* a microorganism that causes infection only under especially favorable conditions (e.g., when defense host mechanisms are not fully functioning)

PART 2

Practical Applications

CHAPTER 11

Practical Bacteriology

WATER and wastewater treatment specialists should be constantly aware of safety precautions whenever they work with living microorganisms, which may include pathogens. This personal safety awareness is especially important when performing laboratory analysis of microorganisms. The following safety rules for laboratory work deserve special attention.

1. While working in the laboratory, wear a clean laboratory coat or smock.
2. Label all chemical containers indicating the chemical name and date of preparation and/or container opening.
3. Check the labels on chemical containers before using to ensure that the proper chemicals are selected for use.
4. Dispose of chemicals safely. Never throw chemicals into common trash cans.
5. Read and learn the directions for each chemical's use and safety. This information is found on the chemical's label or Material Safety Data Sheet (MSDS).
6. Follow directions carefully. *Never* mix chemicals randomly or indiscriminately.
7. Immediately clean up/contain chemical spills according to the directions on the chemical's MSDS.
8. Avoid personal contact with chemicals and microorganisms by using proper personal protective equipment (PPE).
9. Wear gloves for the chemical or microorganisms being handled.

10. Ensure that protective gloves are free of cracks and holes and that they fit properly before handling chemicals or microorganisms.
11. Do not place fingers into the mouth, nose, ears, or eyes while handling chemicals or microorganisms.
12. Wash hands with a disinfectant soap after handling chemicals or microorganisms.
13. Dispose of all contaminated wastes by placing them into designated containers.
14. Avoid contaminating the environment with aerosols containing live microorganisms.
15. Provide positive ventilation to laboratory work areas.
16. Properly dispose of all broken, chipped, or cracked glassware.
17. Use a suction bulb to pipette microorganisms, chemicals, or wastewater. *Never* use the mouth to suck up a fluid or chemical in a pipette.
18. Ensure that the laboratory is equipped with a functional emergency eyewash.
19. Report all spillages and accidents, promptly, to the immediate supervisor.

CULTURE MEDIA

Most of the laboratory media discussed in this section are bacteriological media. Before beginning a discussion of the types, preparation, and use of media, it will be helpful to define or explain some key terms commonly used in bacteriology.

DEFINITION OF KEY TERMS

1. *Agar*—a gelling agent obtained from seaweed that is used to solidify media
2. *Aseptic technique*—the manipulation of a microbial culture in a way that prevents contamination
3. *Autoclaving*—sterilizing with steam under pressure in a gas-tight chamber

4. *Broth*—may refer to liquid media of any type, but when not qualified, commonly refers to nutrient broth

5. *Contaminants*—all unwanted microorganisms

6. *Culture*—a medium containing organisms that are growing or have grown on or within that medium

7. *Differential media*—media that distinguish between different groups of bacteria

8. *Incubated*—medium that is kept under controlled conditions of temperature and humidity for a specific period of time

9. *Incubator*—a thermostatically controlled cabinet where incubation is carried out

10. *Inoculation*—the process of adding the inoculum to the medium

11. *Inoculum*—adding a small amount of material containing living cells to a sterile medium

12. *Medium* (pl. media)—solid or liquid preparation made specifically for the growth of bacteria

13. *Petri dish*—the vessel into which the medium in liquid state is poured and allowed to set

14. *Plate*—a Petri dish containing the solid medium

15. *Pure culture*—a culture consisting of only one type of microorganism

16. *Selective media*—media that favor the growth of particular microorganisms. An example is MacConkey's broth, which inhibits nonenteric bacteria but does not affect the growth of enteric species.

17. *Sterile medium*—a medium containing no living organisms

TYPES OF MEDIA

Microorganisms require nutrients as well as a favorable environment for growth. Therefore, in order to culture microorganisms, the culture medium must contain those nutrients essential for growth of the given microorganism. Additionally, this medium must provide a suitable environment for growth. Important environmental factors such as pH, oxygen, temperature, and osmotic pressure must be considered when selecting media.

In regard to the nutrients that microorganisms need to grow in or on media, the following factors must be taken into account: carbon, water, nitrogen, mineral, energy, growth factors, and pH. The role of each of these factors is discussed in the following sections.

Carbon

Autotrophs can utilize the carbon in carbon dioxide for synthesis of cell materials. *Heterotrophs,* on the other hand, must utilize organic compounds for the carbon that is needed to stimulate their growth. Because of the wide diversity of microorganisms, their need for organic carbon may vary widely with their complexity.

Water

A cell's protoplasm consists of approximately 80% water. Single-celled organisms exist in a water environment contiguous with their internal water supply. This is the case because the cell must be able to pass molecules into and out of its structure freely. Moreover, for critical enzymatically controlled chemical reactions to occur with the cell, water must be present. An important point to keep in mind when preparing media is to use distilled water only. Ordinary tap water, for example, could introduce harmful chemical contaminants into the medium.

Nitrogen

Heterotrophs get their nitrogen from amino acids and protein compounds such as peptones and peptides. Autotrophic organisms can utilize inorganic sources of nitrogen.

Minerals

Metallic elements such as sodium, calcium, magnesium, potassium, iron, copper, phosphorus, and zinc are required by all microorganisms for normal growth. The amounts required are small.

Energy

With the exception of media used for *photoautotrophic* microorganisms, which are able to utilize solar energy, all other media must contain substances that microorganisms can convert to energy. The *chemoautotrophs* can oxidize inorganic substances for their energy needs. Two substances that can provide the essential energy compounds they need are nitrate and nitrite. The *chemoheterotrophs* (most bacteria), on the other hand, require an organic source of energy such as amino acids.

Growth Factors

Along with carbon and nitrogen, some microorganisms require additional materials (growth factors) for growth. These growth factors may include amino acids and vitamins. Although certain microorganisms are capable of synthesizing their own required vitamins, most must obtain them from their nutrient medium.

pH

For some microorganisms, the pH of medium must be within certain limits (the pH range of medium affects enzymatic reactions). Generally, optimal growth of bacteria occurs around pH 7. Other microorganisms grow at other pH levels; thus, levels should be adjusted accordingly.

Essentially, the culture media used in water and wastewater treatment microbiological analysis are of two forms: solid (agar) or liquid (broth). Solid media are made by adding a solidifying agent (agar) to a liquid medium. A good solidifying agent does not liquefy at room temperature. Moreover, a good solidifying agent should not inhibit bacterial growth but should not be utilized by microorganisms. Easy isolation of bacteria is possible on solid media because they allow for growth at a single location. When attempting to isolate microorganisms from mixed colonies it is important to develop colonies on the surface of the medium. Examples of solid media that are commonly used for growing surface colonies of bacteria are blood agar, nutrient agar, and Sabouraud's agar. Blood agar contains defibrinated blood and is used for culture of particular parasites that attack man and other animals. Nutrient agar is a general purpose agar with peptone, salt, and beef and yeast extract added. Sabouraud's agar is generally used for the culture of various fungi pathogenic organisms in man and animals; it is composed of glucose, salt, peptone, and various other solutions (Atlas & Parks, 1993; Singleton & Sainsbury, 1994).

Liquid media such as citrate broth, glucose broth, nutrient broth, and others are also used as general culturing media. These media are commonly used for fermentation studies and for the propagation of large numbers of microorganisms. For example, liquid media can be used for enumeration of cells by the most probable number (MPN) counting technique (McKinney, 1962).

Examples of media used for water and wastewater microbiological analysis are given in Table 11.1.

Table 11.1. Common media used in water/wastewater analysis.

Medium	Used in Water Analysis	Used in Wastewater Analysis
A 1 Broth		XXX
Acetamide agar	XXX	
Actinomycete isolation agar	XXX	
AT5N		XXX
Rose bengal Glucose peptone Agar		XXX
Azide broth	XXX	XXX (Sewage)
Azide citrate broth	XXX	XXX (Sewage)
Azotobacter vinelandii medium	XXX	
BCP azide broth	XXX	XXX
Brilliant green bile broth with MUG	XXX	XXX
Lab-Lemco broth		XXX
MacConkey agar without crystal violet	XXX	XXX

MEDIA CATEGORIES

Before discussing the different general purpose categories of media used for the cultivation of microorganisms, it should be pointed out that the composition of media can be categorized as two general types: *synthetic* and *nonsynthetic* media. Synthetic media are prepared to exact specifications, that is, all components are known. Moreover, the known components of synthetic media are purified and precisely defined and can be easily reproduced. Nonsynthetic media, such as nutrient broth, contain ingredients that are added in exact portions, like the yeast and beef extract portions of nutrient broth.

General purpose media are used to grow many different microorganisms. However, when specially fortified media are needed to encourage the growth of particular heterotrophs, two categories are used: *selective* and *differential* media. Selective media favor the growth of particular microorganisms in or on them because of (1) the presence of inhibitory substances (usually salt or crystal violet) that prevent certain types of organisms from growing on them or (2) the absence of critical nutrients that make it unfavorable to most, but not all, organisms. Differential media cause some bacteria to take on a different appearance from other species, permitting tentative identification of individual microorganisms (e.g., MacConkey's Agar; Singleton & Sainsbury, 1994).

PREPARATION OF MEDIA

For most routine bacteriological work, media preparation has been greatly simplified by the use of dehydrated media. Prior to the development of dehydrated media, laboratory workers had to prepare their own media from raw materials. This was an exacting and time-consuming activity. Today, thanks to the development of dehydrated media, all that is required is to dissolve a measured amount of dehydrated (powdered) medium in water, steam to dissolve the agar, adjust the pH, distribute into tubes or flasks, and sterilize.

For uses such as streaking (to be discussed later) excess surface moisture on newly made plates should be allowed to evaporate; this can be accomplished by placing the plate in the incubator at about 38°C for 30 minutes. Keep in mind that some media contain constituents that may be destroyed if heated at 37–38°C. To avoid destroying these heat-sensitive constituents, it is best to steam them by *autoclaving*.

ASEPTIC TECHNIQUE

Before the water or wastewater specialist proceeds to use media to obtain cultures of microorganisms for growth and study, all unwanted or contaminant organisms must be excluded; that is, the media must be sterilized and maintained in a sterile condition (as was described previously). Inoculating a sterile medium with a pure culture of microorganisms without outside contamination is known as *aseptic technique*. During bacteriological procedures, instruments and materials being used must be protected from contamination by microbes that are always present in the environment. Preuse procedures should include using aseptic technique on all instruments, media, vessels, etc.

In preventing contamination of media, instruments, and materials, air-borne contaminants are the most common problem, since air always contains populations of microbes. When opening containers, care and caution must be exercised to prevent contaminant-laden air from entering. Keeping containers at an angle so that most of the opening is not exposed to air is one way of reducing the possibility of contamination. The preferred method for manipulation of cultures is to carry out the operation in a dust-free space absent of air currents generated by fans, HVAC systems, and open doorways. When transferring a culture from one vessel of medium to another with an inoculating needle, wire, or loop, sterilize these instruments in open flame to redness *before* and *after* making a transfer (aseptic transfer).

In order to further reduce the risk of contamination, the work area (laboratory bench, etc.) should be treated with a disinfectant. When environmental air contamination might be a

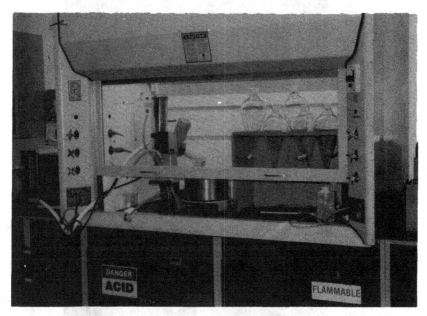

FIGURE 11.1 *Combination fume hood/sterile cabinet. Air flow patterns with the cabinet can be directed to create a barrier to contaminants.*

problem, it is best to perform the bacteriological culture work in a fume hood (see Figure 11.1).

Laboratory fume hoods or sterile cabinets supply filtered air to the work surface; thus, the work can be performed in a contaminant-free environment. When working with pathogenic microbes, a more sophisticated sterile cabinet must be used. These sterile cabinets must be gas tight, fed filtered air, and entry by the technician for manipulation purposes must only be made by using attached arm-length rubber gloves (glove-box technique).

During the actual transfer of cells, a loop or straight wire is commonly used. A loop is generally made of nichrome wire, has a closed loop at the end, and is held in a handle of about 11 cm in length. A straight wire is about 6 cm in length and is attached to a metal handle (see Figure 11.2).

The loop and/or straight wire are used to pick up small quantities of growth from a bacterial colony by bringing the wire loop, or the tip of the straight wire, into contact with the colony.

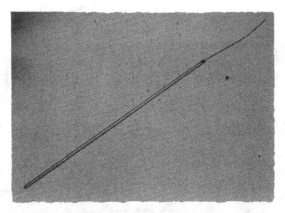

FIGURE 11.2 *A straight wire used for inoculating culture media.*

FIGURE 11.3 *Flaming a straight wire before use.*

The solid or liquid material carried by a loop or straight wire can be used to inoculate either a solid or a liquid medium. Remember to always use the aseptic technique; that is, flame the loop or straight wire immediately before and after use (see Figure 11.3).

INOCULATION TECHNIQUES

Once the inoculating loop or wire has been sterilized, the next step in culturing a microorganism is inoculation of the medium with the specimen. Inoculation and transfer techniques are important to master because they are normally part of many different bacteriological procedures such as isolation of pure cultures, preparation of samples for microscopic examination, and transfer of microbial growth from one medium to another.

Several procedural steps are taken in a typical tube inoculation. The tube containing the inoculum and one or more tubes with sterile media are held together in one hand. The thumb keeps the tubes in proper position while the other hand holds the inoculating wire (see Figure 11.4). The wire is passed

FIGURE 11.4 Standard tube inoculation procedure.

through a flame (flamed) and remains there until redness is obtained. The cotton plugs or tube stoppers are then grasped by the fingers or the hand holding the inoculating tool. The mouths of these tubes must be flamed. When inoculum is obtained and introduced into the proper tubes, the mouth openings of the tubes must be passed through the flame again. Then the plugs or stoppers are inserted into the tubes. Don't forget, the inoculating tool must be flamed to redness again (see Figure 11.3).

CULTURAL CHARACTERISTICS

After inoculation and incubation, microbial growth will normally be apparent. Certain cultural characteristics, such as type of growth on agar slants, pellicle formation in broth cultures, pigmentation, and colonial appearance of organisms can be observed on plates (see Figure 11.5). These cultural characteristics are often useful in describing a particular bacterial species. Moreover, these growth patterns often give information about an organism's relation to air.

Culture characteristics are different for each type of medium used; this is quite apparent when viewing broth media, agar slant media, or streak plate media. Cells may not grow at all in broth media (see Figure 11.6). When growth does occur, it may be dispersed in a manner that makes the media look cloudy or turbid. In other broth cultures, organisms may grow in a film across the surface called a *pellicle*, which may be heavy or light and membranous. Some organisms grow around the broth-air interface and form what resembles a ring (see Figure 11.6).

A slant culture is prepared by melting a small amount of solid medium in a tube and allowing it to solidify in a slanted position. When pure cultures are streaked on the surface of the slant culture, the cultural growth characteristics of the organism can be observed on the slant [see Figure 11.7(a) and (b)]. It should be pointed out that the terms used in Figure 11.7(b) for agar slant culture patterns are outmoded and not often used today because so many variables affect the pattern (e.g., amount of surface moisture, the medium, inoculum size, and the amount of agar included).

Circular   Filamentous   Irregular   Rhizoid   Spindle   Punctiform

ELEVATION

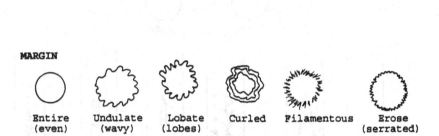

Flat   Raised   Convex   Umbonate   Pulvinate

MARGIN

Entire (even)   Undulate (wavy)   Lobate (lobes)   Curled   Filamentous   Erose (serrated)

FIGURE 11.5 *Cultural characteristics of bacteria seen with the naked eye in relation to their form, elevation, and appearance at the margin; these characteristics are often useful in describing and identifying a particular bacterial species. The general shape of the margin or edge can be determined by looking down at the top of the colony. The nature of the colony elevation is apparent when viewed from the side.*

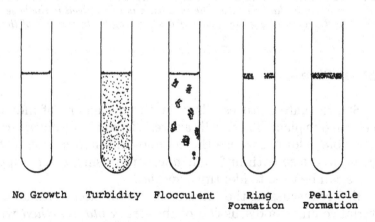

No Growth   Turbidity   Flocculent   Ring Formation   Pellicle Formation

FIGURE 11.6 *Selected cultural characteristics of broth cultures.*

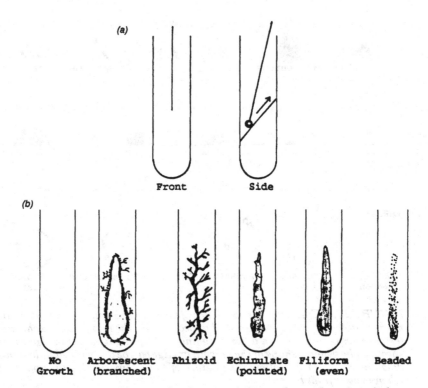

FIGURE 11.7 *(a) Front and side view of slant culture; (b) cultural charac-teristics of agar slant culture. The inoculum is introduced to the base of the slant and drawn along the surface from this region to the end of the agar slant.*

Streak Plate

Streak plate cultures allow for the placement of individual cells in one place. These cells are restricted to one place forming a visible colony. Because the colonies differ in shape, texture, color, and size with different microorganisms, colony appear-ance can be used to identify a species.

Before discussing the cultural characteristics of isolated bac-terial colonies, a discussion of the *streak plate method* will be presented. The streak plate method allows the water or wastewater specialist the opportunity to view and study the cul-tural, morphological, and physiological characteristics of an in-dividual microorganism (a pure culture).

The first step in the streak plate method begins with preparing the Petri dishes (plates) for inoculation. This is accomplished by selecting the proper medium for the isolation and growth of the specific microbe intended. The chosen medium is then placed in dishes (plates) as shown in Figure 11.8.

The streaking process involves inoculation of the surface of a solid medium so that individual colonies of microorganisms form on at least part of the surface. To begin the streak, the microbial mixture is transferred to the edge of a dried agar plate with a loop and then is swiped back and forth (streaked) over the surface as shown in Figure 11.9. At some point in the process, single cells drop from the loop as it is swiped along the agar surface and develop into separate colonies. Performing the streaking process correctly requires skill gained with experience. After growing cultured colonies on the streak plate, the naked eye can see variations in bacterial colony morphology. These morphological characteristics are shown in Figure 11.5.

FIGURE 11.8 *Preparing Petri dishes for inoculation.*

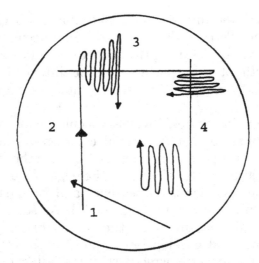

FIGURE 11.9 *A pattern of inoculation used in one method of streaking.*

COUNTING BACTERIA

Changes in number of cells can be used to measure growth. It must be pointed out, however, that when attempting to determine the total numbers of bacteria in water or wastewater samples, no single medium, temperature, or pH is universally ideal. This is the case because water organisms have great variability in physiological needs.

There are several methods for counting cell numbers. The two methods that will be covered in this text are the *total cell count* and *viable cell count* methods. The total number of living and dead cells in a given volume of sample is called the total cell count. Usually, total cell counts refer to single-celled organisms like bacteria, yeasts, or spores. The total cell count is accomplished by counting under the microscope. The count is done either on samples dried on slides or on samples in liquid. Cultures grown in Petri dishes with sterilized membrane grids (Figure 11.10) can be counted using a stereo microscope as shown in Figure 11.11 (Singleton, 1992).

The total cell count in a liquid (broth) sample can be estimated by direct counting in a counting chamber. The cell count

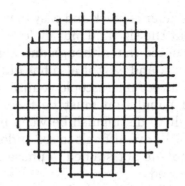

FIGURE 11.10 *Sterilized membrane used for cell counting.*

can be accomplished in a counting chamber by using a grid marked on the surface of a glass slide with squares of known size. The number of cells per unit area of grid can be counted under the microscope giving a measure of the number of cells per chamber volume. Conversion of this value to the number of cells per mass of suspension is accomplished by multiplying a conversion factor based on the volume of the chamber sample.

FIGURE 11.11 *Using a stereo microscope to count cells.*

A counting chamber is quick and easy to use; it also can give information about the size and morphology of microorganisms.

It should be noted that although direct microscopic counting is relatively quick and easy, it also has disadvantages. The disadvantages of direct microscopic counting of cells include (1) The microbial population must be fairly large for accuracy. (2) It is difficult to distinguish between living and dead cells. (3) Very small cells are difficult to see under the microscope. (4) When the sample is not stained, a phase-contrast type microscope must be used.

The *viable cell count* method is used in medical, food, and aquatic microbiology to count the number of living single-celled organisms in a sample. Cultures of organisms are diluted in a liquid medium and placed in an environment that allows them to produce some type of visual effect. The actual count of bacteria is of bacterial colonies. For this reason, the viable count is often called the *colony count*. In this type of counting procedure the assumption is that each viable cell will produce one colony. Thus, the number of cells in the sample is estimated from the number of colonies that develop on or within the medium.

Plate Counts

There are two ways of performing plate counts: the *spread plate* method and the *pour plate* method. In the spread plate method, the sample is pipetted onto the surface of a sterile agar plate of less than 0.1 ml and is spread evenly over the surface of the plate. The plate is then incubated. After incubation, viable cells are estimated from the number of colonies, from the volume of sample (inoculum) used, and from the degree to which the sample had been diluted. If a sample is suspected of containing many cells, it can be diluted in tenfold steps. Several tenfold dilutions of the sample are commonly used (see Figure 11.12); at least one dilution will give a countable number of colonies.

In the pour plate method of plate counting, a known volume of inoculum is mixed with a molten agar-based medium. Because the sample is mixed with molten agar medium, a larger volume can be used than with the spread plate. When incubated, colonies develop within (subsurface colonies) and on

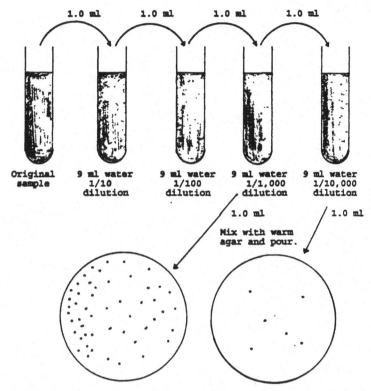

FIGURE 11.12 *Procedure used in pour plate method for viable count using serial dilutions of the sample.*

(surface colonies) the medium, and the viable count is calculated as in the spread plate method.

SUMMARY OF KEY TERMS

- *Aseptic technique:* technique carried out to prevent the introduction of contaminating organisms
- *Selective medium:* growth medium containing a component that will inhibit the growth of certain microorganisms and either enhance or not affect the growth of other organisms

CHAPTER 12

Coliforms

WATER containing a large number of microorganisms may be perfectly safe to drink. The important consideration, from a microbiological standpoint, is the kinds of microorganisms that are present. Water treatment specialists are primarily concerned with bacteria in public water supplies. Wastewater specialists, on the other hand, are concerned with bacteria and other microorganisms that might be returned with treated effluent dumped into a receiving body. Public law requires that owners of all public water supplies collect water samples and deliver them to a certified laboratory for bacteriological examination at least monthly. The number of samples to be collected is usually mandated by federal standards, which generally require that one sample per month be collected for each 1,000 persons served by the water works (see Table 12.1).

Bacteriological tests on water and wastewater determine whether pathogenic organisms are present by testing for certain indicator organisms. In practice, these indicator organisms are the focus of testing because it is not practical to identify specific disease-producing organisms present in water; to check water for each pathogenic agent would be difficult and time consuming.

Among the organisms that are present in large numbers in the intestinal tract of humans and warm-blooded animals are a variety of *coliform* bacteria, *Clostridium perfringens*, and fecal streptococcus. When these organisms are present in water, it indicates that the water has received contamination of an intestinal origin. The *Clostridium perfringens* and streptococcus bac-

163

Table 12.1. Minimum bacteriological sampling
requirements for public water supplies.

| Population Served | | Samples per |
From	To	Month
25	1,000	1
1,001	2,500	2
2,501	3,300	3
3,301	4,100	4
4,101	4,900	5
4,901	5,800	6
5,801	6,700	7
6,701	7,600	8
7,601	8,500	9
8,501	12,900	10
12,901	17,200	15
17,201	21,500	20

Source: Federal Register, June 29, 1989

teria are pathogenic but are difficult to isolate, test for, and can cause serious health hazards for the laboratory technicians who may be exposed to them. Therefore, in the United States, the total coliform group of bacteria is used as indicator organisms for measuring and thus determining the bacteriological quality of drinking water.

The coliform group is defined as all the aerobic and facultatively aerobic, gram-negative, non-spore-forming, rod-shaped bacteria that ferment lactose gas in 24–48 hours at 35°C. As applied to the *membrane filter technique* (to be discussed later), the term applies to a group of gram-negative, non-spore forming, rod-shaped bacteria that develop a red colony with a green metallic sheen within 24 hours at 35°C on medium containing lactose.

Coliform bacteria are members of the family Enterobacteriaceae. They include the genera *Escherichia* and *Aerobacter*. Although originally thought to be entirely of fecal origin, many species have been found in soil. The *total coliform test* checks for coliforms from all origins.

Coliforms are the indicator organisms of choice for testing biological contamination in water supplies for several reasons. (1) An absence of coliforms is usually indicative of the absence of pathogenic microorganisms, since coliforms are generally more persistent in the environment. (2) When an intestinal pathogen is discovered, coliforms are usually present in much greater numbers. (3) Coliforms are easy to isolate using relatively unsophisticated equipment. (4) Coliforms are safe to work with in the laboratory. (5) The density of coliforms present is roughly proportional to the amount of fecal contamination present. (6) Results of laboratory testing of coliforms can be obtained as quickly as 24 hours after sampling.

In the previous instance of explaining why coliforms are the indicator organisms of choice in determining the presence/absence of biological contamination in water it was stated that an absence of coliforms usually indicates the absence of pathogenic microorganisms. The key word in this statement is *usually*. The point to remember is that the presence or absence of coliforms *may* indicate that pathogenic microorganisms are present or absent; likewise, it is important to remember that there are exceptions. That is, the absence of coliforms in water does not always indicate a clean bill of health. As a case in point, consider the following study conducted by Blostein (1991). In July 1989 an outbreak of shigellosis occurred among visitors to a recreational park in Oakland County, Michigan. *Note:* Shigellosis is an acute bacterial infection of the bowel, characterized by abdominal pain, diarrhea, and fever. It is generally transmitted by contact with the feces of individuals infected with a pathogenic species of bacteria of the genus *Shigella* (*Mosby's*, 1990).

An epidemiologic investigation revealed an association between illness and swimming in a pond at the park, especially for those who had submerged their heads underwater. No other factors were identified epidemiologically. A total of sixty-five cases were identified; nine were culture confirmed, all *Shigella sonnei*. Fecal coliform counts were taken on several water samples shortly after the outbreak and were found satisfactory. Cultures of water samples were negative for *Shigella* spp. Additionally, an inspection of the park's sewage disposal system found all equipment in proper working order with no evidence of sewage contamination from this potential source. No other commercial or residential sources of sewage contamination were found near the pond. Investigators concluded that *Shigella* contamination of the pond by a swimmer or swimmers on one or more occasions was a strong possibility. Several factors supported this conclusion. There was, for example, an elevated incidence of *Shigella sonnei* in the community during the two months prior to the outbreaks. There was also an increased use of the pond by swimmers. Moreover, warm water and air temperatures with inadequate water exchange in the pond were contributors.

This important report is one of few documented cases on the growing number of incidents that have been associated with recreational use of water. This example also demonstrates that even when bacteriological tests indicate the absence of coliforms from a particular sampling site, the possibility of being contaminated by pathogenic fecal-type microorganisms at that particular site is not eliminated.

The total coliform group includes four genera in the Enterobacteriaceae family: *Escherichia, Klebsiella, Citobactor,* and *Enterobacter.* Some of the total coliform species may be pathogenic, although most are not. Of this group, the *Escherichia coli* species appears to be most representative of fecal contamination. As a matter of public record, most research conducted in the area of indicators of water contamination has centered on the genus *Escherichia* (especially the species *E. coli*); this becomes obvious when one looks at the volumes of data that have been collected on this particular indicator. However, researchers, such as Ewing et al. (1981), point out that there is little documentation based upon examination of large numbers of cultures and tests of *Escherichia coli.* Furthermore, according to Salyers and Whitt (1994), even though *E. coli* is an "important cause of human and animal disease worldwide," little effort has been made to study *E. coli* as a pathogen (p. 191). The point is that much more research on *E. coli* is needed.

Ewing et al. point out several additional problems associated with using *Escherichia coli* as the primary indicator organism for water contamination. Other problems arise from the facts that (1) certain organisms found in water that do not represent fecal pollution are able to ferment lactose, (2) research has shown that *E. coli* is not a single species as was previously thought, and (3) *E. coli* identical to that found in humans is also found in the intestinal tract of other animals.

Notwithstanding the previously mentioned problems with using *E. coli* as the primary indicator of fecal-based water contamination, if fecal coliform or *E. coli* bacteria are found in water, there should be an immediate concern that the water may be carrying disease-causing organisms. When it comes to laboratory analysis, more definitive test data are possible when tests are performed for coliforms *E. coli*, and *Streptococcus faecalis*. When used together, *E. coli* and *S. faecalis* are good indicators of sewage or fecal contamination.

SUMMARY OF KEY TERMS

- *Coliform:* gram-negative rods, including *Escherichia coli* and similar species that normally inhabit the colon of humans and animals
- *Coliform, Total:* a group of gram-negative, aerobic to facultative anaerobic, non-spore-forming, rod-shaped bacteria that ferment lactose at 35°C producing gas and acid within 48 hours. Using the membrane technique, they develop a red colony with a green, metallic sheen within 24 hours at 35°C on an Endo-type medium containing lactose. They consist of genera *Escherichia, Enterobacter, Citobactor,* and *Klebsiella*
- *Fecal coliform:* the thermotolerant form of *E. coli* and *Klebsiella* that produce a blue colony on a membrane filter when cultured on M-FC medium, and incubated at 45°C for 24 hours
- *Fecal streptococcus:* a number of species such as *S. faecalis, S. faecium,* and others
- *Lactose:* a carbohydrate sometimes known as milk sugar

CHAPTER 13

Bacteriological Examination of Water

AS pointed out earlier, water may contain a large quantity of bacteria and still be safe to drink or to dump as treated wastewater effluent into a receiving body. Moreover, from a bacteriological standpoint, the important consideration is the kinds of microorganisms that are present. Water sources such as lakes or streams that contain multitudes of autotrophs and saprotrophic heterotrophs are potable as long as pathogens are not present. For humans it is the intestinal pathogens such as those that cause cholera, typhoid fever, and bacillary dysentery that are of prime importance. These intestinal pathogens are borne by human fecal material that is carried away by water in sewage systems. In many cases, these sewage systems dump their waste streams into lakes and streams. Obviously, this situation presents a serious sanitation problem. Thus, constant testing of municipal water and wastewater systems for the presence of fecal microorganisms is essential for the maintenance of water purity.

The preceding chapter pointed out that the preferred method to test for the presence of microorganisms in water calls for the use of selective nutritional media with indicator systems to grow organisms into colonies that are visible, easy to identify, and countable. Testing for the pathogen itself is a difficult undertaking for two reasons. Current test methods are too complex, and direct handling of the pathogen puts the handler at risk; instead, nonpathogenic indicator organisms are tested for. Indicator organisms should not only be safe to handle but also survive in the water to be tested. Moreover, the indicator organisms must have a population density relative to the degree of pathogenic contamination. At the same time, the indicator organisms must be able to outlive the pathogens, yet disappear rapidly following the pathogens' demise. The bottom line is that safe water should not contain any of the indicators. Finally, the indicator test used must be able to easily produce quantitative results on a regular basis. The three commonly used indicator tests/techniques are the *multiple tube fermentation* (MTF) procedure, the *membrane filter* method, and the *presence-absence* (P-A) procedure.

COLIFORM TESTING

It is not likely that water or wastewater specialists working at small water or wastewater treatment facilities will actually perform the laboratory tests for coliform bacteria. However, water and wastewater specialists should be familiar with the basic concepts of these tests so they can interpret laboratory results.

Before discussing the three common indicator methods used to determine the bacteriological quality of water, it must be pointed out that microbiological testing of water is a five-step process: sampling, filtration, culturing, incubation, and examination. The water sample must represent (i.e., be a representative sample) the water source's microbial flora. Under the U.S. EPA's Total Coliform Rule each test requires a minimum standard sample portion of 100 ml.

The three tests mentioned earlier have been developed to detect total coliform bacteria, fecal coliforms, and *E. coli* based on characteristics that define each group. Currently, three different groups of tests for total coliforms are acceptable for monitoring for compliance under the bacteriological standards of the Safe Drinking Water Act.

Multiple Tube Fermentation (MTF) Procedure

The first group of tests is known as the multiple tube fermentation procedure and is composed of three steps. In the presumptive step a series of 9 or 12 tubes of lactose broth are inoculated with measured amounts of water sample. The samples are incubated at 35°C for 48 hours, after which the laboratory technician determines if there was any fermentation of lactose (indicator of coliform presence) by looking for gas in the inverted tube. If gas is seen in any of the lactose broths, coliforms are presumed to be present in the water sample. This test is also used to determine the *most probable number* (MPN) of coliform present per 100 ml of water. *Note:* Most Probable Number (MPN)—If the bacteria are randomly distributed in the water sample and a portion of the sample is progressively diluted with sterile buffered water, a point will be reached where no bacteria are transferred to the next dilution. Statistical tables have been constructed to estimate the probable density of bacteria in the original sample.

Step two of the MTF procedure is known as the confirmed test. In this test, plates of a more selective agar are inoculated from positive (gas-producing) tubes to see if the organisms that are producing the gas are gram-negative (a coliform characteristic). The selective media inhibits the growth of gram-positive bacteria and causes colonies of coliforms to be distinguishable from noncoliforms. Depending on the type of selective agar used, small colonies with dark centers (nucleated colonies) or reddish colonies may be apparent. The presence of either of these coliform-like colonies confirms the presence of a lactose-fermenting, gram-negative bacterium.

The final step in the MTF procedure is called the completed test. This step determines if the isolate from the agar plates truly matches the definition of a coliform. The media used in this step include a nutrient agar slant and a Durham tube of lactose broth. If gas is produced in the lactose tube and a slide from the agar slant reveals that a gram-negative non-spore-forming rod is present, it is certain that a coliform is present.

The completion of these three steps with positive results establishes that coliforms are present; however, there is no certainty that *E. coli* is the coliform present. The complete process requires at least four days of incubation and transfers.

Membrane Filter (MF) Procedure

Since the published acceptance of the MF procedure in the 11th edition of *Standard Methods*, work carried out by numerous investigators in microbiology has shown that MF procedure is generally superior to the MPN procedure for routine analysis. The following advantages of the MF procedure over the original MPN procedure are cited by the U.S. Environmental Protection Agency (1978):

1. Results can be obtained in approximately 24 hours, compared to up to 96 hours for the MTF procedure.
2. Larger, more representative samples of water can be analyzed routinely with membrane filters.
3. Numerical results from membrane filters have much greater reproducibility than is expected with the MTF procedure.
4. The equipment required is user friendly. A large number of samples can be examined with minimum requirements for laboratory space, equipment, and supplies.

Additionally, in its publication, "Microbiological Methods for Monitoring the Environment" (U.S. EPA, 1978), the EPA states that the MF procedure is preferred because it permits analysis of large sample volumes in reduced analysis time.

Because the results of bacteriological water testing often can be invalidated by delays in processing, *Standard Methods* recommends field testing when there would otherwise be more than a six-hour delay between sampling and processing. The MF procedure is currently the only approved method that easily lends itself to field testing.

In the membrane filter procedure the 100 ml water sample is passed through a 150-micrometer-thick membrane filter. The filter with its trapped bacteria (those larger than 0.47 micrometers) is transferred to the surface of a solid medium or to an absorption pad containing the desired liquid medium. Use of the proper medium allows the rapid detection of total coliforms, fecal coliforms, or fecal streptococci by the presence of their characteristic colonies. For example, when the filter is examined for bacterial colonies, those formed by coliforms grown on nutri-ent-rich M-Endo broth and inoculated for 24 hours at 35°C will have a golden-green, metallic sheen when observed under fluo-rescent light. If coliforms are detected on the filter, they must be verified by transferring the bacterial colonies to a tube con-taining lactose broth medium (called LTB) and an inverted gas-collection tube—following the steps used in the MTF proce-dure.

The membrane filter procedure will provide a count of the coliform density in the original sample (each colony represents one organism in a sample); a count of one colony or more is re-ported as "present" for regulatory purposes. Colonies must be verified by incubating them in lactose broth to see if they pro-duce gas.

The entire Coliform, Fecal-Membrane Filtration procedure devised from information taken from the Rules and Regula-tions section of the *Federal Register,* the EPA's "Microbiological Methods of Monitoring the Environment" and *Standard Methods* is provided in its entirety in the Appendix.

Presence-Absence (P-A) Procedures

Before discussing the third group of tests (presence-absence or P-A procedures), some background information will be help-ful.

The MPN procedure or the MF technique have traditionally been the accepted bacteriological testing procedures used in the examination of water samples to determine whether the quality of water is acceptable for human consumption and other domestic uses (APHA, 1989). Interest in presence-absence test methods for determining the microbiological quality of drinking water has steadily increased in recent years. On 31 December 1990, when the Total Coliform Rule promulgated by the U.S. Environmental Protection Agency became effective, the P-A methodology became not only officially available but also readily available as manufac-turers jumped on the marketing bandwagon (*Federal Register,* 1989). The new regulation changed the manner of reporting total coliforms from numbers per 100 ml to the presence or absence of total coliforms in 100 ml of sample.

The fundamental impact felt by the water and wastewater specialist (whose task it is to perform bacteriological examination of water and wastewater) is the availability of a simplified test for detecting coliforms and fecal coliforms. This test is a modification of the MPN procedure and depends on a change in the pH in the sample container when coliform bacteria have fermented lactose. A 100-ml or larger water sample is incubated in a single culture bottle with a triple-strength broth containing lactose broth, lauryl tryptose broth, and bromcresol purple indicator. The P-A test assumes that no coliforms should be present in 100 ml of drinking water. A positive test results in the production of acid (the purple dye in the lactose broth turns yellow) and constitutes a presumptive test requiring confirmation.

Several manufacturers have devised their own P-A type test kits for sale and commercial use. In addition, a test for both coliform and *E. coli*, the Colilert defined substrate test, is available for use and will be discussed later. Several studies have been conducted to determine the effectiveness and reliability of the various P-A test kits that are currently being marketed. For example, a study conducted by Clark and El-Shaarawi (1993) evaluated several commercial presence-absence test kits during a six-month period in 1990. The test kits evaluated included the Colilert, Coliquik, Hach Disposable, and the Hach Vial test. The standard used for comparison with these kits was the Ontario Ministry of the Environment (MOE) P-A test. The product evaluations were based on the general principles of the multiple-tube fermentation technique. Each week, a surface water sample was diluted and inoculated into twenty-five 99-ml dilution blanks for each of three dilutions. The inoculated dilution blanks from each dilution series were randomly sorted into sets of five. Three of these sets were inoculated into the P-A test kits or vice versa, as required. The other two sets were passed through membrane filters, and one set of five membrane filters was placed onto m-Endo agar LES to give replicate total coliform counts and the other set was placed onto m-TEC agar to give replicate fecal coliform results. A statistical analysis of the results was performed. The comparative test results showed that three of the four commercial products gave only fair levels of recovery when the data were compared with the data from MOE P-A tests and membrane filter tests. P-A bottles showing positive results after 18 hours of incubation that were subcultured immediately in ECMUG tubes frequently could be confirmed as containing total coliforms, fecal coliforms, or *Escherichia coli* after 6 hours of incubation; thus, the total incubation time was only 24 hours. This study confirmed not only that the commercially available P-A test kits give results in 24 hours, as is claimed by the manufacturers but also demonstrated an accuracy that was in line with the results normally obtained with the MOE P-A test.

About the same time that the original P-A test kits were first introduced and evaluated by various research teams, an alternative test procedure for detecting total coliforms and *Escherichia coli* was being investigated. This new test was put on the market by Access Analytical Systems under the name of Colilert. A similar product, known as Coliquik, was developed and marketed by the Hach Company. Several studies have been conducted to determine whether the Colilert and Coliquik tests detected total coliforms and *E. coli* as well as the *Standard Methods* MPN and P-A procedures do.

One such comparison study was conducted by Clark et al. (1991). In their study, the researchers focused on determining the performance capabilities of two commercial 4-methylumbelliferyl-β-D-glucuronide (Colilert and Coliquik test kits) preparations in the detection of *Escherichia coli* from water samples. More than eighty samples were collected from a treated water reservoir, and thirty-two samples were collected from untreated surface water. The results of the comparison study indicated that there was a statistically significant difference between the two commercial preparations compared with the *Standard Methods* membrane filtration fecal coliform method for the detection of *E. coli* from treated water samples. However, there was no difference between the two methods and the MFC test for *E. coli* detection from the untreated surface water samples. The disagreement between the two commercial products and the MFC method was primarily caused by the occurrence of false-negative results with the two commercial products. Research data indicated that the occurrence of false-negative samples could be attributed to impaired substrate specificity and sensitivity of the two tests for *E. coli* detection. There was no apparent relationship between the occurrence of false-negative results and heterotrophic plate counts in samples. In their final report, the researchers recommended that before tests for the detection of *E. coli* are incorporated into the *Federal Register* they should be tested and evaluated extensively by a number of laboratories throughout the nation.

A few months after their original study, a follow-up study by the same research team compared membrane filtration with Colilert and Coliquik test kits in detecting total coliforms in drinking water (Olson et al., 1991). This study used the presence-absence format with the test kits against the *Standard Methods* membrane filtration technique to determine whether differences existed in total coliform detection. The results indicated that both the Colilert and Coliquik test kits were acceptable methods for total coliform detection.

Results of studies like those just mentioned and of several others indicate that the P-A test kits that are currently available can be used with confidence. Moreover, to test for coliforms and fecal coliforms, a variety of simpler and more specific tests have been developed. Additionally, research is ongoing in the effort to develop even more efficient, precise, and user-friendly test equipment and kits. As a case in point, take the example reported by Drake (1995) where a company in Connecticut is developing a DNA test using the polymerase chain reaction (PCR) to detect the presence of harmful waterborne protozoans known for causing infections to humans. The use of PCR has become quite common in environmental testing. Using PCR in conjunction with DNA tests in surface waters to detect *Giardia* in a rapid manner is another potential time-saving, user-friendly testing technique that could aid water specialists in the performance of their important testing duties.

SUMMARY

In order for water and wastewater specialists to perform their demanding and highly technical duties to ensure that treatment processes operate at optimum performance, a basic understanding of microbiology is essential. For the wastewater specialist who may be responsible for ensuring the correct operation of a biological treatment process that might include activated sludge, trickling filters, and rotating biological contactors (RBCs), some fundamental knowledge of microbiology is important. This is especially the case when one considers that a strong correlation exists between the type of organism that predominates in the activated sludge and the quality of the effluent produced. As an example where knowledge in the fundamentals of microbiology can aid the wastewater specialist in the perfor-

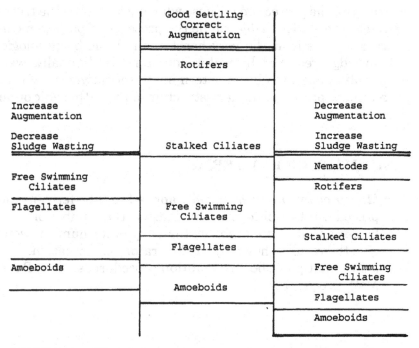

FIGURE 13.1 *The relative predominance of protozoa in the mixed liquor of a biological system as it relates to indicating the degree of biological activity that is taking place. For example, a predominance of lower life forms such as amoebas and flagellates indicates a young sludge condition and is usually associated with poor floc formation and settling. On the other hand, a majority of rotifers and stalked ciliates with nematodes present indicates an old sludge condition. Increased wasting is often recommended to correct this situation. Ideally, a predominance of free-swimming ciliates and stalked ciliates in the mixed liquor indicates that effective settling and, thus, effective treatment is taking place.*

mance of his or her duties, consider the following examples. If, for instance, the plant operator notices that the process is dominated by flagellates, there might also be a corresponding very poor BOD_5 and TSS removal with no floc formation. On the other hand, the same process may be dominated by stalked ciliates where there is excellent BOD_5 and TSS removal with excellent flocculation and clarity. These are operating parameters that can help wastewater operators perform their jobs in a much more efficient and professional manner (see Figure 13.1).

In the case of the water specialist, a basic (in many cases more advanced) knowledge of and training in microbiology is critical to ensure the protection of those who may consume the treated water. The ability to determine the presence of pathogenic microorganisms is at the very heart of the basic bacteriological knowledge required by water operators. Additionally, water specialists must be able to determine the effectiveness of water treatment processes in the destruction of the pathogenic organisms.

SUMMARY OF KEY TERMS

- *Most probable number:* usually applied to tube fermentation procedures for coliforms. This index is the number of bacteria that, more probably than any other number, would give the results shown by the laboratory examination.
- *MTF:* multiple tube fermentation procedure, such as P-A or MPN procedures

PART I: COLIFORM, FECAL-MEMBRANE FILTRATION (FC-MR)

I. DISCUSSION

Fecal coliforms belong to the total coliform group. They are gram-negative, non-spore-forming rods that ferment lactose at 24 ± 2 hours at $44.5 \pm 0.2°C$, with the production of acid; the organisms form blue colonies on mFC broth.

The major species of the fecal coliform group is *Escherichia coli*, indicative of fecal pollution and the possible presence of enteric pathogens. The test is applicable for examination of the following water sources: lakes, public water supplies, runoff, sewage, and wells. This method is not applicable for saline water samples.

II. SAMPLE HANDLING AND PRESERVATION

A. Container type

    1. Samples for fecal coliform analysis are collected in sterile 250-ml glass jars with screw caps. (The caps are lined with rubber and are autoclavable).

Adapted from *Federal Register*, Volume 56, No. 195 (1995); *Federal Register*, Volume 43, No. 209 (1994); EPA (1978), "Microbiological Methods of Monitoring the Environment"; and *Standard Methods* (1992).

2. Sample jars must be filled with sample to within one inch of the top of the jar to allow for adequate mixing of the sample.

B. Interference removal
Sample jars must contain 0.2 ml 10% sodium thiosulfate ($Na_2S_2O_3$) per 250-ml jar to neutralize chlorine and 0.6 ml 15% EDTA per 250-ml jar to chelate metals.

C. Samples must be refrigerated or iced down immediately after collection and be maintained at 1–4°C until analyzed.

D. Holding time
Samples kept at 1–4°C must be analyzed within 6 hours, including collection time. Unrefrigerated samples must be analyzed immediately.

III. APPARATUS AND EQUIPMENT

A. Apparatus
1. Water bath, 44.5 ± 0.2°C
2. Binocular dissecting microscope with a 10–15× objective magnification and a cool fluorescent lamp
3. Hand tally
4. Forceps, with smooth, round tips
5. Sterile Petri dishes, 50 × 12 mm, plastic, with absorbent pads meeting quality control specifications (see Section VII, Quality Control)
6. Sterile membrane filters, 47 mm diameter, 0.45 μm, individually wrapped meeting quality control specifications (see Section VII, Quality Control)
7. Water-proof plastic bags (Whirlpak)
8. Oxford-type transfer pipettor, calibrated quarterly (See Section VII, Quality Control)
9. Sterile disposable tips (5 and 10 ml) for pipettor
10. Sterile, 100 and 50 ml, graduated cylinder covered with aluminum foil
11. Sterile 1-ml and 10-ml pipets

12. Sterile membrane filtration unit, free of scratches and leaks
13. Sample jars, 250 ml, sterilization resistant (see Appendix D, Glassware Preparation)
14. Bunsen burner and gas source
15. Striker or matches
16. Appropriate top-loading balance
17. Analytical balance with a sensitivity of 1.0 mg
18. Hot plate, with magnetic stirrer
19. Magnetic stir bars
20. pH meter accurate to at least 0.1 pH unit
21. Combination pH electrode
22. Autoclave
23. Vacuum source
24. Thermometer, able to measure 44.5°C, graduated in increments of 0.2°C or less, calibrated against NIST (NBS) or NIST traceable thermometers semiannually with NIST corrections marked

IV. REAGENTS

A. mFC broth
B. Phosphate buffered dilution water
C. Sodium thiosulfate ($Na_2S_2O_3$) solution, 10% (w/v)— 1 L. Dissolve 100 g $Na_2S_2O_3$ in approximately 900 ml high-purity water. Dilute to 1 L.
D. Ethylenediamine tetraacetate (EDTA) dihydrate solution 15% (w/v)—1 L
    1. Dissolve 150 g EDTA in approximately 900-ml high-purity water. Dilute to 1 L.
    2. Adjust pH to 6.5 with concentrated sulfuric acid (H_2SO_4).
    Note: Wear eye protection and gloves.
E. 95% ethanol

V. PROCEDURE

A. Preparing the plates
    1. Prepare at least three sterile plates for each sample to be analyzed. If the chlorine residual is less

than 1.50 mg/L or if the samples seem dirtier than normal, prepare an additional plate.

2. Prepare one plate for a blank to be planted at the end of each filtration series. Add an additional plate for a blank after every ten samples planted.

3. Using an Oxford transfer pipettor with a 5 ml tip or a sterile pipet, aseptically dispense approximately 2.1 ml of mFC medium into each plate.

4. Number the plates with a permanent marking pen.

5. Aseptically, pour off the excess broth from the plates, just prior to planting the samples.

B. Planting the sample

1. Record the required information on the worksheet. Include anything unusual about the sample.

2. On a routine basis, plant 1, 10, and 100 ml volumes for each sample. If the chlorine residual is less than 1.50 mg/l, add a 0.1 ml dilution.

a. For a 0.1 ml dilution:

Aseptically transfer 9 ml of buffer water into a sterile tube. Using a 1 ml pipet, transfer 1 ml of shaken sample to the dilution tube to make a 10^{-1} dilution. Plant 1 ml of the 0.1 ml dilution using a sterile 1 ml pipet.

3. One blank sample must be placed at the end of the filtration series to check for contamination. When a newly sterilized filtration unit must be used, plant a blank at the end of the new filtration series. An additional blank must be planted after every ten samples. Plant blanks following the same guidelines as the sample, using 100 ml of buffer water instead of sample.

4. A filtration series is considered interrupted when 30 minutes elapses. After 30 minutes, switch to another sterile filtration unit. Do not switch filtration units in the middle of a sample.

5. Sterilize the forceps by dipping in the alcohol and

placing in a flame. Allow adequate cooling before using the forceps to handle a membrane.

6. Remove the top portion of the filter funnel, place a filter (grid side up) on the lower portion of the filter apparatus using the forceps and remount the top portion of the funnel. Be sure the membrane filter is smooth.

7. When less than 20 ml of sample (diluted or undiluted) is filtered, add a small amount of dilution water (5–10 ml) to the funnel before filtration. This aids in uniform bacterial dispersion.

8. Shake the sample twenty-five times.

9. Dispense the appropriate amount of sample slowly onto the filter. If an Oxford type pipettor is used, release the last drop of sample by depressing the knob further by pulling it away from the finger rest with your thumb as you are pushing down, disengaging the calibration stop.

10. Without touching the sides of the funnel, remove the pipet and discard the pipet tip.

11. If the 100 ml volume will not filter because of high solids or polymer content, discard the filter and replate, splitting the volume onto two plates or four plates, i.e., 50 + 50 ml, or 25 + 25 + 25 + 25 ml. If the 50 or 25 ml will not filter, add a 0.1-ml and four additional 10-ml dilutions and use the 0.1-, 1- and five 10-ml aliquots plate counts to calculate FC/100 ml.

12. If the 100 or 50 ml volumes will filter but the membrane contains an excessive amount of solids which may affect colony growth, split the volumes into two or four subaliquots, i.e., 50 + 50 ml, or 25 + 25 + 25 + 25 ml.

13. If a 25, 50, or 100 ml portion of a sample is planted, measure the volume with a sterile graduated cylinder and carefully pour into the funnel.

14. Turn on the vacuum and allow the sample to filter through.

15. If the funnel leaks, the sample must be planted

again. Change to another sterile funnel before re-planting the sample. Run another blank at the end of the new filtration series. Record on the worksheet.

16. After the sample filters, rinse the inside of the funnel at least three times, using 20–30 ml of buffer water for each rinse.

17. When the water has filtered, flame the forceps, turn off the vacuum, and tilt off the top of the funnel.

18. With the sterilized forceps, remove the membrane without tearing and place the funnel back into the funnel base.

19. Carefully place the membrane (grid side up) on the absorbent pad in the Petri plate with a rolling motion to avoid air bubbles. The membrane surface should be smooth and not torn. To avoid contamination, do not allow the membrane to touch any surface except for the absorbent pad and put the lid back on the Petri plate immediately.

20. Tap the plate on the counter gently to set the membrane. Center the filter, making sure it is not caught in the lid; tap gently on the side of the plate in the direction you want the filter to move.

21. Check for air bubbles under the filter and push any air bubbles out with the sterilized forceps, being careful not to tear the membrane.

22. With a waterproof marker, label the plate with the sample code, date and the volume filtered.

Note: The plates must be in the incubator within 30 minutes after filtration.

C. Incubating the plates

1. When all of the samples and the blanks have been filtered, arrange the plates in stacks of two in the water-proof plastic (whirlpak) bags. Do not put more plates in the bag than can be securely sealed. All of the plates must have the same side up.

2. To seal, hold the whirlpak by the white tabs and

whirl it around. Avoid getting too much air in the bag. Fold the tabs in and then away from the bag so that the wires will not puncture the bag. Place the bag into a second whirlpak bag and seal. Label the bag with date, time in, and technician initials. Record the time in the incubator on the laboratory worksheet.

3. Put the whirlpak on the bottom shelf of the wire rack in the water bath so that the plates are all upside down.

4. Put a lead weight on the top of the rack so that the plates will remain underwater.

5. Incubate in the water bath at 44.5 ± 0.2°C for 24 hours ± 2 hours.

Note: The plates must be in the water bath within 30 minutes after the samples have been filtered and must remain underwater throughout incubation.

D. Reading the plates

1. The plates are read after 24 ± 2 hours of incubation.

2. Fecal coliform colonies are blue, greenish-blue, or clear with blue centers. Nonfecal coliform colonies are gray, buff, or colorless and are not counted.

3. Count the fecal coliform colonies using a dissecting scope on 10× power with a fluorescent light. Pin-point size colonies are not counted unless confirmed (see Part II: Coliform, Fecal-Verification Procedure).

4. Record the number of fecal coliform colonies on each plate on the worksheet.

5. If colonies are too indistinct to obtain an accurate count owing to colony morphology, make a note on the sheet. Use the plate counts from the other volumes to calculate FC/100 ml.

6. If confluent growth occurs (growth covering either the entire area of the membrane or a portion and colonies are not discrete), record the count as "confluent growth with fecal coliforms" if the

growth is blue or greenish-blue. If the confluent growth is not typical of fecal coliforms, then record as "confluent growth without fecal coliforms." If confluent growth only covers a portion of the membrane, make a note on the sheet.

VI. RESULTS

A. Calculation of report values for nonsplit 100-ml volumes. If the 100-ml volume was not split, then proceed as follows. If the 100 ml volume was split, proceed to Section VI, B.

1. Calculation:

$$FC/100 \text{ ml} = \frac{\text{No. of colonies counted}}{\text{Vol. of sample filtered (ml)}} \times 100$$

2. Selecting correct plate counts for calculation of reported values

a. Select the sample quantity that produces a plate count within the desired range of twenty to sixty colonies and calculate the FC/100 ml. See Example 1, Section A, 3.

b. If all of the plates have no colonies, select the largest volume filtered; calculate the FC/100 ml as if there was one colony on the plate. Report as a "<" value. See Example 2, Section A, 3.

c. If all the sample volumes have plate counts < 20, total the number of colonies in all volumes and calculate the FC/100 ml. Remember to include the total volume in the calculation. See Example 3, Section A, 3.

d. If none of the sample volumes have plate counts in the desired range, but some counts are <20 and some are >60 (or TNTC), select the plate with counts closest to the desired range and calculate FC/100 ml. See Example 4, Section A, 3. If more than one sample volume have plate counts equally distant from the desired range, calculate the FC/100 ml for both and average the results for a final reported value. See Example 5, Section A, 3.

e. If more than one volume per sample have plate counts in the 20–60 range, calculate the counts per 100 ml for each volume and average the results. Report the average as the final result. See Example 6, Section A, 3.

f. If all sample volumes have greater than 60 but less than 200 total colonies per plate, or if all plate counts are TNTC (>200 colonies/plate), select the smallest volume filtered and calculate as if there were 60 colonies on the plate. Report result as a ">" value. See Example 7, Section A, 3. On the following day, add a smaller volume (100, 10, 1, and 0.1 ml). Do not delete the other volumes.

g. If a plate(s) exhibits atypical colony morphology that is too indistinct for an accurate count, make a note on the sheet. Use plate counts from the other volumes to calculate the FC/100 ml. See Example 8, Section A, 3.

h. If a plate count for a particular volume cannot be used, use plate counts from the other sample volumes to calculate the FC/100 ml.

Record all results on the worksheet. Be sure to put date, time, and your initials. Submit the worksheet to the chief chemist/microbiologist after the results have been checked by another analyst.

3. Examples

	Volumes Filtered				
	1 ml	10 ml	100 ml	Plate	FC/100
Example	No. of Colonies/Plate			Count/Vol	ml
1	3	32	TNTC	32/10 ml	320
2	0	0	0	1/100 ml	<1
3	5	12	18	35/111 ml	32
4	12	18	65	18/10 ml	180

(continued)

	Volumes Filtered					FC/100
	1 ml	10 ml	100 ml	Plate		FC/100
Example	No. of Colonies/Plate			Count/Vol		ml
5	3	18	62	18/10 ml	180	121
				62/100 ml	62	
6	3	20	59	20/10 ml	200	130
				59/100 ml	59	
7	61	100	199	60/1 ml		>6,000
8	TNTC	TNTC	TNTC	60/1 ml		>6,000
9	1	12	2	13/11 ml		118

B. Calculation of reported values for 100-ml split volumes. When the 100-ml volume was divided into two 50-ml or four 25-ml subaliquots, proceed as follows for calculation of reported values.

  1. Calculation

$$\text{FC/100 ml} = \frac{\text{No. of colonies counted}}{\text{Vol. of sample filtered (ml)}} \times 100$$

  2. Plate counts and calculation of reported values

   a. If all the plates for all the sample volumes have no colonies including the split volume plates, then select the largest volume (which for the 100 ml split volume would be 100 ml) and calculate the FC/100 ml as if there was one colony on the plate. Report as a "<" value. See Example 1, Section B, 3.

   b. If the plates for other volumes have <20 colonies and the plates for the 100 ml split volumes also have <20 colonies, total plate counts for all volumes and calculate the FC/100 ml. (Remember to use the total volume in the calculation.) See Example 2, Section B, 3.

   c. If none of the sample volumes have plate counts within the desired range, but some are <20 and some are >60 (or TNTC), select the volume with plate count closest to the desired

[2]Colony morphology too indistinct for an accurate count.

range to calculate FC/100 ml. See Example 3, Section B, 3. If more than one sample volume have plate counts equally distant from the desired range, calculate the FC/100 ml for both and average the result for a final reported value. See Example 4, Section B, 3.

d. If any of the plates for the 100-ml split volume have colony counts within the 20–60 range, then calculate the FC/100 ml for each subaliquot with colony counts within the range. If more than one volume (including subaliquots) have plate counts within the range, calculate the FC/100 ml for each and average the results for the final reported value. See Example 5, Section B, 3.

e. If all of the plates for the 100-ml split volume as well as plates from other volumes have 20–60 colonies, calculate the FC/100 ml for the 100-ml split volume by adding the colonies. Then calculate the FC/100 ml for each of the other volumes meeting the criteria. Average all of the results to obtain a final reported value. See Example 6, Section B, 3.

f. If all plates have >60 but <200 total colonies per plate, or if all plate counts are TNTC (>200 colonies/plate), select the smallest volume filtered and calculate as if there were 60 colonies on the plate. Report result as a ">" value. See Example 7, Section B, 3. On the following day, add a smaller volume (100, 10, 1 and 0.1 ml). Do not delete the other volumes.

g. If a plate(s) exhibits atypical colony morphology which is too indistinct for an accurate count, make a note on the sheet. Use plate counts from the other volumes to calculate the FC/100 ml. See Example 8, Section B, 3.

h. If a plate count for a particular volume cannot be used, use plate counts from the other sample volumes to calculate the FC/100 ml.

Record all results on the worksheet. Be sure to indicate date, time, and initials. Submit to chief chemist/microbiologist after the results have been checked by another analyst.

3. Examples

Exam-ple	Volumes Filtered			Plate Count/ Volume		FC/100 ml
	1 ml	10 ml	100 ml Split			
	No. of Colonies/Plate					
1	0	0	0, 0, 0, 0	1/100 ml		<1
2	0	2	18, 16, 17, 15	68/111 ml		61
3	0	18	62, 61	61/50 ml		122
4	0	18	62, 63	18/10 ml	180	
				62/50 ml	124 /	152
5	0	8	18, 21, 23, 17	21/25 ml	84	
				23/25 ml	92 /	88
6	3	20	59, 56, 57, 60	20/10 ml	200	216
				232/100 ml	232	
7			TNTC, TNTC	60/1 ml		>6,000
	TNTC	TNTC	TNTC, TNTC	60/1 ml		>6,000
8	1	12	2,2	13/11 ml		118

VII. QUALITY CONTROL

A. Apparatus

1. Membrane filters

a. Use only 0.45 μm-pore diameter, 47-mm diameter membranes with grid lines that have been certified by the manufacturer to exhibit full retention of the organisms, stability in use, no chemical leeching that affects growth and development of bacteria, satisfactory speed of filtration, no significant influence on media pH, and no increase in confluent growth. Bacterial growth should not be influenced along the grid lines. For presterilized membranes, the manufacturer should certify that the sterilization technique has neither induced toxicity nor altered the chemical or physical properties of the membranes. Main-

[2]Colony morphology too indistinct for an accurate count.

tain a log of the membrane lot numbers used, including manufacturer certifications.

b. Store membranes in a location free from humidity and temperature extremes. Do not store more than a year's supply.

2. Petri dishes

Disposable, sterile, moisture tight plastic Petri dishes (50 × 12 mm) with tight-fitting lids and sterile absorbent pads. The dish bottoms should be flat and large enough so that the absorbent pad lies flat. The dishes are to be certified by the manufacturer as moisture tight and free from residual growth suppressive effects. Maintain a log of lot numbers used, along with manufacturer certifications.

3. Absorbent pads

Sterile, with sufficient thickness to absorb 1.8–2.2 ml of media, and certified by the manufacturer to release less than 1-mg total acidity (calculated as $CaCO_3$).

4. Filtration units

a. Constructed of autoclavable plastic. Consists of a seamless funnel fastened by a magnetic device and designed so that the membrane will be held securely in place on the porous plate without damage and so that all fluid will pass through during filtration.

b. The top and bottom portions of the magnetic filter funnel are wrapped and sterilized separately.

c. A new sterile filtration unit must be used at the beginning of each filtration series. A blank must be run before removing filtration unit after samples have been filtered in order to test for cross-contamination.

d. The filtration unit must be free from leaks and scratches.

5. Forceps

Round tipped, free from corrugations on the inner side. Forceps are sterilized before each use by dipping in 95% ethyl or absolute methyl alcohol and flaming.

6. Dissecting microscope

   Colonies are counted with a magnification 10–15 diameters and a light source that gives a maximum sheen discernment. Use a binocular widefield dissecting microscope with a small, cool-white, fluorescent lamp.

7. Thermometer

   Measures 44.5°C, has 0.2-degree increments, and is calibrated at least semiannually against a NIST or NIST traceable thermometer with NIST corrections indicated.

8. Oxford transfer pipettor

   Check for proper volume setting before each use. Pipettor should be calibrated monthly.

B. Prepared media

1. Media lot comparison

   Before a new lot of media is used, perform a lot comparison.

2. Verify fecal coliform colonies and test for false negative colonies for 10 fecal coliform colonies/month (see Part II: Coliform, Fecal-Verification Procedure).

3. Records

   Maintain a bound book with the following records:

   a. Date of media preparation and preparer's initials
   b. Media type and lot number
   c. Amount weighed and volume prepared
   d. Sterilization time and temperature
   e. pH measurements and adjustments
   f. Colony verification records
   g. Lot comparison records

C. Sterility

1. Check for sterility of media, membrane filters, dilution and rinse water, glassware, and equipment

once each filtration series by planting 100 ml of buffer water as a blank sample.

2. If contamination is indicated, reject analytical data from samples.

3. All procedures must be carried out in a manner that will neither inhibit desired bacterial growth nor cause contamination.

D. Precision

1. Multiple analysts

a. Once per month, each analyst performing analysis throughout the month should count typical colonies on the same membrane from a positive sample. Verify colonies (see Part II: Coliform, Fecal-Verification Procedure) and compare counts to the verified count.

b. Once per month, each analyst performing analysis throughout the month should perform parallel analysis on a positive sample.

VIII. DATA VALIDATION

The data for the fecal coliform–membrane filtration procedure should be labeled invalid if any of the following instances occur:

A. The coliform water incubator is not within range of $44.5 \pm 0.2°C$ during the incubation period.

B. The samples were collected in a nonsterile container.

C. The sample jars did not contain sodium thiosulfate to neutralize chlorine and EDTA to chelate metals where needed.

D. Samples were not maintained at 1–4°C prior to analysis.

E. Samples were analyzed after the 6-hour holding time had expired.

F. The mFC medium was prepared incorrectly.

G. Sample dilutions were prepared improperly.

H. Culture plates were improperly labeled (e.g., improper sample source, improper sample volume).

I. Improper medium was selected for the analysis.

J. Nonsterile pipettes, pipet tips, graduated cylinders sample plates, membranes, absorbent pads or other type glassware were used for sample analysis.

K. Sample plates were not incubated for the appropriate time of 24 ± 2 hours.

L. Water leaked into the plates during incubation, resulting in questionable plate counts.

M. Water depth in the coliform water incubator was not sufficient to submerge the sample plates.

N. Fecal colonies on the plates were improperly counted.

O. Results were recorded improperly, yielding questionable results.

P. Dehydrated opened media older than 6 months was used for media preparation.

PART II: COLIFORM, FECAL-VERIFICATION PROCEDURE

I. DISCUSSION

Verification of the membrane filter test for fecal coliforms establishes the validity of colony differentiation by the blue color and provides supporting evidence of colony interpretation. The verification procedure corresponds to the fecal coliform MPN (EC medium) test. Verification must be run on at least 10 colonies each month, provided there are 10 positive results.

II. SAMPLE HANDLING

Blue colonies are taken from samples that have been analyzed by the fecal coliform membrane filtration procedure.

III. APPARATUS AND EQUIPMENT

A. Apparatus
  1. Sterile inoculating loops
  2. Culture tubes, 75 × 10 mm

3. Culture tubes, 150 × 20 mm
4. Incubator, 35 ± 0.5°C
5. Water bath, 44.5 ± 0.2°C
6. Autoclave
7. pH meter
8. Combination pH electrode
9. Top-loading balance
10. Transfer pipettor
11. Disposable tips for transfer pipettor
12. Hot plate with magnetic stirrer
13. Magnetic stir bars

IV. REAGENTS

A. Lauryl tryptose broth
B. EC medium

V. PROCEDURE

A. With a sterile loop, pick the center of a well-isolated blue colony. Inoculate a tube of LTB. Repeat for a total of ten colonies (ten tubes).
B. Incubate the tubes for 24–48 hours at 35 ± 0.5°C. Read the tubes at 24 and 48 hours; note the positive tubes. A positive tube is indicated by the presence of gas and turbidity.
C. Confirm the gas-positive LTB tubes at 24 and 48 hours by inoculating an EC tube with a loopful of growth from the LTB broth tubes.
D. Incubate the EC tubes in a water bath for 24 ± 2 hours at 0.2°C water bath. Production of gas and turbidity indicates a positive test. Maintain the water depth in the incubator sufficient to immerse the tubes to the upper level of the medium.

VI. RESULTS

A. Culture tubes that produce gas and turbidity in EC tubes are interpreted as verified fecal coliform colonies.
B. Results for both the LTB and EC broth must be re-

corded in the log book, along with the sample type and date as follows:

Example: 10/26/95 Fecal Coliform Colony Confirmation: CE 102695FC0400

	Tube Inoculated with a Fecal Coliform Colony									
	1	2	3	4	5	6	7	8	9	10
24-hour LTB results	+	–	+	–	+	–	+	–	+	+
48-hour LTB results		+			+					
EC results	+	+	+	+	+	+	+	+	+	+

C. A percent verification can be determined for any colony-validation test:

$$\frac{\text{No. of colonies meeting verification test}}{\text{No. of colonies subjected to verification}} \times 100 = \%$$

Example: Twenty blue colonies on M-FC medium were subjected to verification studies. Eighteen of these colonies proved to be fecal coliforms according to provision of the test:

$$\text{Percent verification} = \frac{18}{20} \times 100 = 90\%$$

1. A percent verification figure can be applied to the direct test results to determine the verified fecal coliform count per 100 ml.

$$\frac{\text{Percent verification}}{100} \times \text{Count per 100 ml}$$

$$= \text{Verified fecal coliform count}$$

Example: For a given sample, by the M-FC test, the fecal coliform count was found to be 42,000 organisms per 100 ml. Supplemental studies on selected colonies showed 92% verification.

$$\text{Verified fecal coliform count} = \frac{92}{100} \times 42{,}000$$

$$= 38{,}640$$

Rounding Off = 39,000 fecal coliforms per 100 ml

The analyst is cautioned not to apply percentage of verification determined on one sample to the other samples.

D. Changing reported results following verification
1. If a fecal coliform colony is picked and carried throughout the verification process and the result is negative, record the changed colony count for that specific sample volume only and recalculate the FC/100 ml for that sample only.
2. Change the value and be sure that the correct FC/100 ml value was recorded.

Abeles, R. H., Frey, P. A., & Jencks, W. P. (1992). *Biochemistry*. Boston: Jones & Bartlett Publishers.

Adams, V. D. (1991). *Water & Wastewater Examination Manual*. Chelsea, Michigan: Lewis Publishers, Inc.

American Public Health Association. (1989). *Standard Methods for the Examination of Water and Wastewater*, 17th ed. American Public Health Association, Washington, D.C.

Arasmith, S. (1993). *Introduction to Small Water Systems*. Albany, Oregon: ACR Publications, Inc.

Atlas, R. M. & Parks, L. C. (eds.) (1993). *Handbook of Microbiological Media*. Boca Raton, Florida: CRC Press.

Badenock, J. (1990). Cryptosporidium *in Water Supplies*. London: HMSO Publications.

Benson, H. H. (1994). *Microbiological Applications*, 6th ed., Dubuque, Iowa: Wm. C. Brown Publishers.

Bergey's Manual of Systematic Bacteriology, 8th edition. (1974). Buchanan, R. E. & Gibbons, N. E. (eds.); Williams & Wilkins.

Bingham, A. K., Jarroll, E. L., Meyer, E. A., & Radulescu, S. (1979). *Introduction of* Giardia *Excystation and the Effect of Temperature on Cyst Viability Compared by Eosin-Exclusion and in vitro Excystation in Waterborne Transmission of Giardiasis*, Jakubowski, J. & Hoff, H. C. (eds.). Washington, DC: United States Environmental Protection Agency, pp. 217–229, EPA-600/9-79-001.

Black, R. E., Dykes, A. C., Anderson, K. E., Wells, J. G., Sinclair, S. P., Gary, G. W., Hatch, M. H., & Gnagarosa, E. J. (1981). Handwashing to Prevent Diarrhea in Day-Care Centers. *Am. J. Epidemilo.*, 113:445–451.

Black-Covilli, L. L. (1992). Basic Environmental Chemistry of Hazardous and Solid Wastes. In P-C. Knowles (ed.), *Fundamentals of Environmental Science and Technology* (pp. 13–30). Rockville, Maryland: Government Institutes, Inc.

Blostein, J. (1991). Shigellosis from Swimming in a Park Pond in Michigan. *Public Health Reports*, 106(3):317–322.

Breslow, R. (1990). *Enzymes: The Machines of Life.* Burlington, NC: Carolina Biological Supply, Co.

Brock, T. D. & Madigan, M. T. (1991). *Biology of Microorganisms.* Englewood Cliffs, New Jersey: Prentice-Hall.

Brodsky, R. E., Spencer, H. C., & Schultz, M. G. (1974). Giardiasis in American Travelers to the Soviet Union. *J. Infect. Dis.,* 130:319–323.

Burrows, W. (1966). *Textbook of Microbiology.* Philadelphia: W. B. Saunders Company.

Centers for Disease Control. (1979). Intestinal Parasite Surveillance, Annual Summary 1978. Atlanta: Centers for Disease Control.

Centers for Disease Control. (1981). Biochemical Characterizations *Escherichia coli* 1981, DHHS Publication No. (CDC) 81-8109. Atlanta, Georgia.

Centers for Disease Control. (1981). Water-Related Disease Outbreaks Annual Summary 1979, DHHS Publication No. (CDC) 81-8385. Atlanta, Georgia.

Centers for Disease Control. (1982). Water-Related Disease Outbreaks Annual Summary 1981, DHHS Publication No. (CDC) 82-8385. Atlanta, Georgia.

Centers for Disease Control. (1984). Water-Related Disease Outbreaks Surveillance, Annual Summary 1983. Atlanta: Centers for Disease Control.

Clark, D. L., Milner, B. B., Stewart, M. H., Wolfe, R. L., & Olson, B. H. (1991). Comparative Study of Commercial 4-Methylumbelliferyl-β-D-Glucuronide Preparation with the *Standard Methods* Membrane Filtration Fecal Coliform Test for the Detection of *Escherichia coli* in Water Supplies. *Applied and Environmental Microbiology,* 57(5):1528–1533.

Clark, J. A. & El-Shaarawi, A. H. (1993). Evaluation of Commercial Presence-Absence Test Kits for Detection of Total Coliforms, *Escherichia coli,* and other Indicator Bacteria. *Applied and Environmental Microbiology,* 59(2):380–388.

De Zuane, J. (1997). *Handbook of Drinking Water Quality.* New York: John Wiley & Sons, Inc.

Dhaliwal, B. S. (1979). *Nocardia amarae* and Activated Sludge Foaming. *Journal Water Pollution Control Federation,* 51:344.

Drake, T. (1995, October) DNA Tests Being Developed for Water-Borne Protozoans. *Water/Engineering & Management,* p. 17.

Ewing, J., Davis, D., & Martin, M. (1981). *Biochemical Characterization:* Escherichia coli. U.S. Department of Health & Human Services (PHS), Atlanta: CDC HHS Pub. No. 81-8109.

Fayer, R., Speer, C. A., & Dubey, J. P. (1997). The General Biology of *Cryptosporidium.* In Cryptosporidium *and Cryptosporidiosis,* Fayer, R. (ed.). Boca Raton, Florida: CRC Press.

Federal Register, Vol. 43, No. 209, Friday, October 26, 1984. Rules and Regulations, p. 43260.

Federal Register, Vol. 54, 1989. Drinking Water: National Primary Drinking Water Regulations; Total Coliforms Proposed Rule. pp. 27544–27567.

Federal Register, Vol. 56, No. 195. Tuesday, October 8, 1991. Rules and Regulations, p. 50753.

Frost, F., Plan, B., & Liechty, B. (1984). *Giardia* Prevalence in Commercially Trapped Mammals. *J. Environ. Health,* 42:245–249.

Haller, E. J. (1995) *Simplified Wastewater Treatment Plant Operation.* Lancaster, PA: Technomic Publishing Company, Inc.

Hickman, C. P., Roberts, L. S. & Hickman, F. M. (1990). *Biology of Animals,* 5th ed. St Louis: Times Mirror/Mosby College Publications.

Jahn, T. L., Bovee, E. C. & Jahn, F. F. (1979). *How to Know the Protozoa.* Dubuque, Iowa: Wm. C. Brown Company Publishers.

Jarroll, E. L., Jr., Bingham, A. K., & Meyer, E. A. (1979). *Giardia* Cyst Destruction: Effectiveness of Six Small-Quantity Water Disinfection Methods. *Am. J. Trop. Med. Hygiene,* 29:8–11.

Jarroll, E. L., Jr., Bingham, A. K., & Meyer, E. A. (1980). Inability of an Iodination Method to Destroy Completely *Giardia* Cysts in Cold Water. *West. J. Med.,* 132:567–569.

Jarroll, E. L., Jr., Bingham, A. K., & Meyer, E. A. (1981). Effect of Chlorine on *Giardia lamblia* Cyst Viability. *Appl. Environ. Microbiol.,* 41:483–487.

Jenkins D., Richard, M. G. & Daigger, G. T. (1984). *Manual on the Causes and Control of Activated Sludge Bulking and Foaming.* Water Research Commission, Republic of South Africa (U.S. Distributor: Ridgeline Press, Lafayette, CA).

Jokipii, L. & Jokipii, A. M. M. (1974). Giardiasis in Travelers: A Prospective Study. *J. Infect. Dis.,* 130:295–299.

Kelley, S. G. & Post, F. J. (1989). *Basic Microbiology Techniques.* Belmont, CA: Star Publishing Company.

Kemmer, F. N. (1979). *Water: The Universal Solvent.* Oak Ridge, Illinois: NALCO Chemical Company.

Keystone, J. S., Karden, S., & Warren, M. R. (1978). Person-to-Person Transmission of *Giardia lamblia* in Day-Care Nurseries. *Can. Med. Assoc. J.,* 119:241–242, 247–248.

Keystone, J. S., Yang, J., Grisdale, D., Harrington, M., Pillow, L., & Andrychuk, R. (1984). Intestinal Parasites in Metropolitan Toronto Day-Care Centres. *Can. J. Assoc. J.,* 131:733–735.

Kittrell, F. W. (1969). *A Practical Guide to Water Quality Studies of Streams.* United States Department of Interior.

Kordon, C. (1993). *The Language of the Cell.* New York: McGraw-Hill, Inc.

Koren, H. (1991). *Handbook of Environmental Health and Safety: Principles and Practices.* Chelsea, Michigan: Lewis Publishers.

Lafferty, P. & Rowe, J. (eds.) (1993). *The Dictionary of Science.* New York: Simon & Schuster.

Laws, E. A. (1993). *Aquatic Pollution: An Introductory Text.* New York: John Wiley & Sons, Inc.

LeChevallier, H. A., LeChevallier, M. P. & Wyszkowski. (1977). *Actinomycetes of Sewage Treatment Plants.* EPA Pub. No. 600/2-77/145.

LeChevallier, M. W., Norton, W. D. & Lee, R. G. (1991). Occurrence of *Giardia* and *Cryptosporidium* spp. in Surface Water Supplies. *Applied and Environmental Microbiology*, 57(9):2610–2616.

Lippy, E. C. & Waltrip, S. C. (1985). Waterborne Disease Outbreaks—1964–1980: A Thirty-Five Year Perspective. In *Giardia lamblia in Water Supplies—Detection, Occurrence, & Removal*. American Water Works Association. Denver, Colorado, pp. 67–74.

McGhee, T. J. (1991). *Water Supply & Sewerage*, 6th ed. New York: McGraw-Hill, Inc.

McKinney, R. E. (1962). *Microbiology for Sanitary Engineers*. New York: McGraw-Hill, Inc.

Metcalf & Eddy (revised by Tchobanoglous, G. & Burton, F. L.) (1991). *Wastewater Engineering, Treatment, Disposal, and Reuse*, 3rd ed. New York: McGraw-Hill, Inc.

Mosby's Medical, Nursing, and Allied Health Dictionary. (1990). St. Louis: C. V. Mosby Company.

Neidhart, F. C. (3d.) (1987) Escherichia coli *and* Salmonella typhimurium—*Cellular and Molecular Biology*. Washington, DC: American Society of Microbiology.

Olomucki, M. (1993). *The Chemistry of Life*. New York: McGraw-Hill.

Olson, B. H., Clark, D. L., Milner, B. B., Stewart, M. H., & Wolfe, R. L. (1991). Total Coliform Detection in Drinking Water: Comparison of Membrane Filtration with Colilert and Coliquik. *Applied and Environmental Microbiology*, 57(5):1535–1539.

Panciera, R. J., Thomassen, R. W., & Garner, R. M. (1971). Cryptosporidial Infection in a Calf. *Vet. Pathol.*, 8:479.

Patterson, D. J. & Hedley, S. (1992). *Free-Living Freshwater Protozoa: A Color Guide*. Boca Raton, Florida: CRC Press, Inc.

Peavy, S., Rowe, D. R. & Tchobanoglous, G. (1985). *Environmental Engineering*. New York: McGraw-Hill, Inc.

Pennack, R. W. (1989). *Fresh-Water Invertebrates of the United States*, 3rd ed. New York: John Wiley & Sons, Inc. ·

Pickering, L. K., Evans, D. G., Dupont, H. L., Vollet, J. J., III, Evans, D. J., Jr. (1981). Diarrhea Caused by *Shigella*, Rotavirus, and *Giardia* in Day-Care Centers: Prospective Study. *J. Pediatr.*, 99:51–56.

Pickering, L. K., Woodward, W. E., Dupont, H. L., & Sullivan, P. (1984). Occurrence of *Giardia lamblia* in Children in Day Care Centers. *J. Pediatr.* 104:522–526.

Pipes, W. O. (1978). Actinomycete Scum Formation in Activated Sludge Processes. *Journal Water Pollution Control Federation*, 50:628.

Prescott, G. W. (1978). *How to Know the Freshwater Algae*. Dubuque, Iowa: Wm. C. Brown Company Publishers.

Prescott, L. M., Harley, J. P. & Klein, D. A. (1993). *Microbiology*, 3rd ed. Dubuque, Iowa: Wm. C. Brown Company Publishers.

Public Law 93-523, Dec. 16, 1974, *Safe Drinking Water Act*.

Rentorff, R. C. (1954). The Experimental Transmission of Human Intestinal

Protozoan Parasites. II. *Giardia lamblia* Cysts Given in Capsules. *Am. J. Hygiene,* 59:209–220.

Richard, M.G. (1986, September). "Filamentous Microorganisms and Activated Sludge in Colorado: Operational Problems, Diagnosis and Remedial Actions." Paper presented at the annual meeting of the Rocky Mountain Section of the American Water Works Association/Water Pollution Control Association, Breckenridge, CO.

Ries, A. A. Vugia, D. J., Beingolea, L., Palacios, A. M., Vasquez, E., Wells, J. G., Garcia, N., Swerdlow, D. L., Pollack, M. & Bean, N. H. (1992) Cholera in Piura, Peru: A Modern Urban Epidemic. *Journal of Infectious Disease,* 166(6):1429–1433.

Salyers, A. A. & Whitt, D. D. (1994). *Bacterial Pathogenesis: A Molecular Approach.* Washington, DC: American Society for Microbiology.

Sealy, D. P. & Schuman, S. H. (1983). Endemic Giardiasis and Day Care. *Pediatrics,* 72:154–158.

Singleton, P. (1992). *Introduction to Bacteria,* 2nd ed. New York: John Wiley & Sons.

Singleton, P. & Sainsbury, D. (1994). *Dictionary of Microbiology and Molecular Biology,* 2nd ed. New York: John Wiley & Sons.

Slavin, D. (1955). *Cryptosporidium meleagridis* (s. nov.). *J. Comp. Pathol.,* 65:262.

Strom, P. F. & Jenkins, D. (1984). Identification and Significance of Filamentous Microorganisms in Activated Sludge. *Journal Water Pollution Control Federation,* 56:52.

Sundstrom, D. W. & Klei, H. E. (1979). *Wastewater Treatment.* Englewood Cliffs, New Jersey: Prentice-Hall, Inc.

Tchobanoglous, G. & Schroeder, E. D. (1987). *Water Quality.* Reading, Massachusetts: Addison-Wesley Publishing Co.

Thomas, L. (1974). *The Lives of a Cell.* New York: Viking Press.

Thomas, L. (1982). *Late Night Thoughts on Listening to Mahler's Ninth Symphony.* New York: Viking Press.

Tyzzer, E. E. (1907). A Sporozoan Found in the Peptic Glands of the Common Mouse. *Proc. Soc. Exp. Biol. Med.*

Tyzzer, E. E. (1912). *Cryptosporidium parvum* (sp. nov.), a *Coccidium* Found in the Small Intestine of the Common Mouse. *Arch. Protistenkd,* 26:394.

Upton, S. J. (1997). *Basic Biology of Cryptosporidium.* Kansas State University.

U.S. Environmental Protection Agency (EPA). (1978). *Microbiological Methods for Monitoring the Environment: 1. Water & Wastes.* EPA- 600/ 8-78-017, EMSL-Cincinnati, Ohio: EPA.

U.S. Environmental Protection Agency (1991). Priority List of Substances which May Require Regulation under the Safe Drinking Water Act. *Federal Register,* 56:1470–1474.

Voutchkov, N. (1995, August). UV Disinfection—An Emerging Technology. *Public Works,* 39–40.

Walsh, J. A. (1981). Estimating the Burden of Illness in the Tropics. In *Tropi-*

cal and Geographic Medicine, Warren, K. S. and Mahmoud, A. F. (eds.). New York: McGraw-Hill, pp. 1073–1085.

Walsh, J. A. & Warren, K. S. (1979). Selective Primary Health Care: An Interim Strategy for Disease Control in Developing Countries. *N. Engl. J. Med.,* 301:974–976.

Wanner, J. (1994). *Activated Sludge Bulking and Foaming Control.* Lancaster, PA: Technomic Publishing Company.

Weller, P. F. (1985). Intestinal Protozoa: Giardiasis. *Scientific American Medicine.*

Wistreich, G. A. & Lechtman, M. D. (1980). *Microbiology,* 3rd ed. New York: Macmillan Publishing Co.

Witkowski, A. & Power, J. (1975). *Enzymes: Nature's Catalysts.* Burlington, North Carolina: Carolina Biological Co.

Printed in the United States
by Baker & Taylor Publisher Services